Excel

YEAR 8

Problem Solving Workbook

ESSENTIAL skills

Get the Results You Want!

Allyn Jones

Reprinted 2017, 2018, 2020, 2022, 2023

Reviewed in 2024 for the NSW Curriculum and Australian Curriculum Version 9.0 changes

Reprinted 2025

ISBN 978 1 74125 435 8

Pascal Press
PO Box 250
Glebe NSW 2037
www.pascalpress.com.au

Publisher: Vivienne Joannou
Project editor: Rosemary Peers
Edited by Rosemary Peers
Answers checked by Peter Little
Cover, page design and typesetting by DiZign Pty Ltd
Printed by Vivar Printing/Green Giant Press

Introduction

This book has been specifically written for the **Australian Curriculum**.

The Australian Curriculum (Mathematics) covers six areas:

- Number
- Algebra
- Measurement
- Space
- Statistics
- Probability.

Students are encouraged to develop proficiency with mathematical concepts, skills, procedures and processes, and to use them to demonstrate mastery in mathematics as they solve problems.

This book provides strategies for students when applying mathematical concepts in a variety of routine and non-routine problems of increasing difficulty.

Please go to page 9 for Contents

Please go to page 9 for Contents

HOW TO USE

KEY SKILL

11 Financial mathematics: Profit and loss A

HINTS

- A profit, or a loss, is the difference between the cost price and the selling price.
  A profit is made when selling price > cost price.
  A loss is made when selling price < cost price.
- A profit is made when the cost price is increased to give the selling price.
  A loss is made when the cost price is reduced to give the selling price.
- To write a fraction as a percentage you multiply by 100%.

Reminder!

- Read the question carefully to identify what needs to be found.
- Re-read the question when you've finished your answer to make sure that the question has been answered and your solution makes sense.

Examples

At a garage sale, an antique dealer purchases a chair for $40 and later sells it for $65. What is the profit as a percentage of the cost price when the chair is sold?

Solution

$$\text{Profit} = 65 - 40 = 25$$

$$\text{Profit \%} = \frac{25}{40} \times 100\% = 62.5$$

∴ the dealer makes a profit of 62.5%.

FOCUS on ...

1. The question: Asks you to find the profit made.
2. The information: Gives you the cost price and the selling price of the chair.
3. Your working: You need to find the profit.
   Profit = Selling Price − Cost Price
   = 65 − 40
   = 25
   Write this profit as a percentage of the cost price.
   $\text{Profit \%} = \frac{25}{40} \times 100\%$
   = 62.5
4. Your answer: Make sure you write the correct units: The dealer makes a profit of 62.5%.

A car was originally purchased for $37 500. After 5 years the owner sold it at a loss of 72% on his purchase price. What was the selling price of the car?

Solution

Loss = 72% of 37 500
= 0.72 × 37 500
= 27 000

∴ the loss was $27 000.

Selling price = Cost price − Loss
= 37 500 − 27 000
= 10 500

∴ the car was sold for $10 500.

FOCUS on ...

1. The question: Asks you to find the selling price.
2. The information: Gives you the cost price and the loss as a percentage.
3. Your working: You need to find the amount of loss by finding the percentage of the cost price.
   Loss = 72% of 37 500
   = 0.72 × 37 500
   = 27 000
   ∴ the loss was $27 000.
   The loss is subtracted from the cost price to find the selling price.
   Selling price = Cost price − Loss
   = 37 500 − 27 000
   = 10 500
4. Your answer: Make sure you write the correct units: The car was sold for $10 500.

32 *Excel* Year 8 Problem Solving Workbook

THIS BOOK

Key skill

- Each unit focuses on one key skill from each of four sub-strands of the curriculum, e.g. Financial mathematics: Profit and loss A in the Number sub-strand.

Hints

- Students are given hints to help them solve the problems in the unit.
- These hints may also remind students of rules covered earlier in the book.

Reminder

- Students are given a reminder to always follow certain steps when they begin and end each question.

Examples

- Two examples are provided for each unit with worked solutions that students can follow.

Focus on ...

- A step-by-step method is provided for each example.
- These step-by-step methods are then used to solve the questions in the **Now try these!** section.
- Explanations and tips are provided in the methods to help students in their understanding and to encourage them to develop logical processes when solving problems in this unit.

A step-by-step guide to problem solving

Step 1: Focus on the question

- Read the question at least twice to make sure that you understand what you are being asked to solve.

Step 2: Focus on the information

- Concentrate on the relevant information. In some questions you may be given unnecessary information which will not be useful.
- Decide what you are looking for.

Step 3: Focus on your working

- There are many problem-solving strategies that you can use, including:
  - looking for a simpler problem of the same type
  - using easier numbers
  - working systematically
  - using the guess-and-check method
  - eliminating possibilities
  - drawing a picture or a diagram
  - drawing up a table
  - looking for a pattern
  - working backwards.

Step 4: Focus on your answer

- Make sure you write the correct units (e.g. minutes, metres) and ask yourself these questions:
  - Does the answer make sense? Is it a realistic answer?
  - Have I answered the question I was asked?

Now try these!

- All the questions in this section have been carefully written so that the exact same steps in solving them replicate the steps in the examples.
- The CHALLENGE question is a more challenging question where students will need to use slightly different steps to solve it.

Now try these!

1. A painting is purchased for $120 000 and 4 years later is sold for $186 000. What is the profit as a percentage of the cost price?

2. A family buys an eight-week-old german shepherd pup for $800. After 2 years the family move into rental accommodation and have to sell the dog for a loss of 45%. At what price do the family sell the dog?

3. Carl buys a motor bike for $660. He spends $240 on repairs and sells it for $1200. What is his percentage profit?

4. Don purchased a box of apples. He bought them at the rate of 5 apples for $4 and sold them at 4 apples for $5. What was his profit expressed as a percentage of the cost price?

5. Mitch bought and sold a camera lens. The ratio of his cost price to selling price was 4 : 5. Later he sold a camera tripod for the same percentage profit. If the tripod had originally cost $120, what was the selling price of the tripod?

6. CHALLENGE The manager of Nikki's Nuts bought some nuts from a producer. He purchased 30 kg of nuts at $4 per kilogram and another 20 kg of nuts at $5 per kilogram. If the nuts are to be mixed and sold to make a 75% profit, what will be the sale price per kilogram?

Answers page 132

Excel Year 8 Problem Solving Workbook 33

Revision Tests

- Pairs of revision tests appear regularly throughout the book to test students' understanding.
- Each pair has a test at both an Average and Challenging level of difficulty.

REVISION TEST 3 Level of difficulty—Average

1. At the gym, Sarah spent 40% of the time on the treadmill, 35% on the weights and the remainder in the pool. If she swam for 15 minutes, how long was Sarah on the treadmill?

2. At a '15%-off end-of-season sale', Matt bought a pair of footy boots. If he paid $102 for the boots, what was the original price?

3. A block of land in 2005 was valued at $190 000. By 2010 the value had dropped by 4.5%, but 5 years later had risen in price by 25%. What was the value of the land in 2015?

4. On his 40th birthday Tom purchased a motor bike for $25 000. After Tom injured his leg in an accident he sold the bike at a loss of 35%. At what price did Tom sell his bike?

5. Everything in Norm's electrical store is priced at 40% above cost price. A television is priced at $840. If the store discounts prices by 25%, what profit will be made on the sale of the television?

6. A book originally priced at $12 is discounted by 20%. If GST is set at 10%, what is the reduction in the amount of GST?

Answers page 133

38 *Excel* Year 8 Problem Solving Workbook

Quick answers

- Students can quickly mark their work by referring to the quick answers in **bold** type.

Worked solutions

- Each question has a worked solution so that students can check incorrect answers or alternative ways to achieve the answer.

Worked solutions

NUMBER

Key Skill 1 Integers (pages 10–11)

1. **22°**
Range $= 9 - -13$
$= 9 + 13$
$= 22$
$\therefore$ the temperature range was 22°.

2. **43**
Result $= 14 \times 5 + 8 \times (-3) + 3 \times (-1)$
$= 70 - 24 - 3$
$= 43$
$\therefore$ Yun scored 43.

3. **208°**
Difference $= -2 - -210$
$= -2 + 210$
$= 208$
$\therefore$ the difference is 208°.

4. **−17°**
Temperature $= -1 + 10 \times (-1) + 3 \times (-2)$
$= -1 - 10 - 6$
$= -17$
$\therefore$ the temperature is −17°.

5. **−10**
Result $= 2 \times 3 + 3 \times (-3) + 1 \times (-5) + 2 \times (-1)$
$= 6 - 9 - 5 - 2$
$= -10$
$\therefore$ Amanda scored −10.

6. **−40 °F**
$F = \frac{9}{5}C + 32$
$= \frac{9}{5} \times (-40) + 32$
$= -72 + 32$
$= -40$
$\therefore$ the temperature is −40 °F.

Key Skill 2 Unitary method A (pages 12–13)

1. **6 h**
Time for 1 bricklayer $= 8 \times 3$
$= 24$
Time for 4 bricklayers $= 24 \div 4$
$= 6$
$\therefore$ it would take 6 h.

2. **3 days**
Time for 1 dog $= 4 \times 3$
$= 12$
Time for 4 dogs $= 12 \div 4$
$= 3$
$\therefore$ a bag would last 3 days.

3. **18 h**
Time for 1 woman $= 6 \times 6$
$= 36$
Time for 2 women $= 36 \div 2$
$= 18$
$\therefore$ the job would take 18 h.

4. **64 machines**
Machines for 1 day $= 8 \times 4$
$= 32$
Machines for $\frac{1}{2}$ day $= 32 \div \frac{1}{2}$
$= 64$
$\therefore$ 64 machines are needed.

5. **18**
Total food-days for 1 soldier $= 72 \times 20$
$= 1440$
Total food-days for 80 soldiers $= 1440 \div 80$
$= 18$
$\therefore$ the food will last another 18 days.

6. **$10\frac{1}{2}$ days**
Days for 1 man $= 6 \times 8$
$= 48$
Days remaining $= 48 - 3 \times 6$
$= 30$
Days remaining for 4 men $= 30 \div 4$
$= 7\frac{1}{2}$
Total time $= 7\frac{1}{2} + 3$
$= 10\frac{1}{2}$
$\therefore$ it took a total of $10\frac{1}{2}$ days for the job.

Key Skill 3 Unitary method B (pages 14–15)

1. **2 h**
LCM of 3 and 6: 6
Kirra: fences in 6 h = 2
Bradie: fences in 6 h = 1
Kirra and Bradie: fences in 6 h = 3
As $6 \div 3 = 2$, then it will take 2 h.

2. **18 min 45 s**
LCM of 30 and 50: 150
Pipe A: tanks in 150 min = 5
Pipe B: tanks in 150 min = 3

126 *Excel* Year 8 Problem Solving Workbook

Contents

Number

Algebra

Measurement

Statistics and Probability

KEY SKILL

1 Integers

HINTS

- The number line is useful for adding or subtracting:

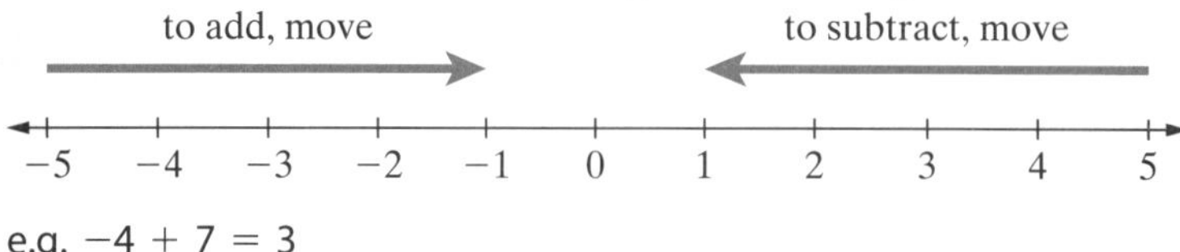

e.g. $-4 + 7 = 3$

- The rules when adding or subtracting directed numbers are:

$+ + = +$ $\quad + - = -$ $\quad - + = -$ $\quad - - = +$

e.g. $-3 - (-2) = -3 + 2$
$= -1$

- The rules when multiplying or dividing directed numbers are:

$+ \times + = +$ $\quad + \times - = -$ $\quad - \times + = -$ $\quad - \times - = +$

$+ \div + = +$ $\quad + \div - = -$ $\quad - \div + = -$ $\quad - \div - = +$

e.g. $12 \div (-2) = -6$

Reminder!

- Read the question carefully to identify what needs to be found
- Re-read the question when you've finished your answer to make sure that the question has been answered and your solution makes sense.

Examples

Two friends stayed at a coral reef resort. Lee went parasailing at a height of 96 m. At the same time, Dawson was scuba diving directly below Lee at a depth of 37 m. Find the distance between them.

Solution

Distance $= 96 - (-37)$
$= 96 + 37$
$= 133$

$\therefore$ the friends are 133 m apart.

FOCUS on...

1. **The question:** Asks you to find the distance between the two friends.
2. **The information:** Gives you the height above the surface and the depth below the surface.
3. **Your working:** You need to subtract the two distances. The depth below the surface is written as −37. Use the rule that subtracting negative is the same as adding (−− = +).
   Distance $= 96 - (-37)$
   $= 96 + 37$
   $= 96 + 4 + 33$
   $= 133$
4. **Your answer:** Make sure you write the correct units: They are 133 m apart.

In a test a correct answer gains 3 marks, an incorrect answer scores −2 and an unanswered question scores 0. In a test with 20 questions, Peta answers 12 questions correctly and 5 answers incorrectly. If she left the other questions unanswered, what was her result?

Solution

Total $= 12 \times 3 + 5 \times (-2) + 3 \times 0$
$= 26$

$\therefore$ Peta scored 26 marks.

FOCUS on...

1. **The question:** Asks you to find the total score on the test.
2. **The information:** Gives you the marks for correct, incorrect and unanswered questions.
3. **Your working:** Find the number of questions not answered and then write an expression involving positive 3 for each correct answer and −2 for each incorrect answer.
   Unanswered $= 20 - (12 + 5) = 3$
   Total $= 12 \times 3 + 5 \times (-2) + 3 \times 0$
   $= 36 - 10$
   $= 26$
4. **Your answer:** Make sure you write the correct units: Peta scored 26 marks.

Now try these!

1. A town's minimum temperature was −13 °C, and later that day the maximum was 9 °C. What was the range in temperature?

2. In a test a correct answer gains 5 marks, an incorrect answer scores −3 and an unanswered question scores −1. In a 25-question test, Yun answers 14 questions correctly and 8 answers incorrectly. If the other questions were not answered, what was her result?

3. Nitrogen freezes at −210 °C and sea water freezes at −2 °C. What is the difference between the two temperatures?

4. The temperature in a ski village at 6 pm is −1 °C. Every hour until 4 am the temperature drops 1 °C, and then 2 °C each hour until 7 am. What is the temperature at 7 am?

5. In a certain game of cards the losing players need to calculate the total points of the cards remaining in their hand. A heart is worth (+3), diamond (−1), club (−3) and a spade (−5). What is Amanda's total score if she has 2 hearts, 3 clubs, a spade and two diamonds?

6. CHALLENGE To change a temperature from Celsius (C°) to Fahrenheit (F°), the formula $F = \frac{9}{5}C + 32$ is used. What is the temperature in Fahrenheit if it is −40 °C.

Answers page 126

KEY SKILL

2 Unitary method A

HINTS

- The unitary method finds the value of one unit which is then used to get the answer. This 'unit' could be, for example, the time it takes for one person to do something,
  e.g. If 3 people take 4 hours to finish a job, how long will it take 1 person?

  $$\text{Time for 1 person} = 4 \times 3 = 12$$

  ∴ it will take 1 person 12 hours.

  Now, if a job takes 1 person 12 hours to complete, how long will it take 2 people?

  $$\text{Time for 2 people} = 12 \div 2 = 6$$

  ∴ it will take 2 people 6 hours.

Reminder!

- Read the question carefully to identify what needs to be found
- Re-read the question when you've finished your answer to make sure that the question has been answered and your solution makes sense

Examples

A tanker uses 3 hoses to fill a container in 20 minutes. How long would it take the tanker to fill the container using 5 hoses?

Solution

$$\text{Minutes for 1 hose} = 20 \times 3 = 60$$

$$\text{Minutes for 5 hoses} = 60 \div 5 = 12$$

∴ it would take 12 minutes.

FOCUS on ...

1. **The question:** Asks you to find the time to fill a container using 5 hoses.
2. **The information:** Gives you the time taken to fill the container using 3 hoses.
3. **Your working:** You first need to find the time using 1 hose so multiply the time by 3. Then divide this answer by 5 to get the time for 5 hoses.

   $$\text{Minutes for 3 hoses} = 20$$

   $$\text{Minutes for 1 hose} = 20 \times 3 = 60$$

   $$\text{Minutes for 5 hoses} = 60 \div 5 = 12$$

4. **Your answer:** Make sure you write the correct units: The tanker takes 12 minutes to fill the container.

Cassie has 4 cats and a bag of cat food lasts 6 days. If she was given 2 more cats to mind, how long would a bag of cat food last if all the cats ate the same amount each day?

Solution

$$\text{Days for 1 cat} = 6 \times 4 = 24$$

$$\text{Days for 6 cats} = 24 \div 6 = 4$$

∴ the food would last 4 days.

FOCUS on ...

1. **The question:** Asks you to find the time the food would last for 6 cats.
2. **The information:** Gives you the number of days the food lasts for 4 cats.
3. **Your working:** You need to multiply by 4 to find the number of days the food would last for 1 cat and then divide this by 6 to find the time it lasts for 6 cats.

   $$\text{Days for 4 cats} = 6$$

   $$\text{Days for 1 cat} = 6 \times 4 = 24$$

   $$\text{Days for 6 cats} = 24 \div 6 = 4$$

4. **Your answer:** Make sure you write the correct units: The cat food would last 4 days.

Now try these!

1. Three bricklayers complete a wall in 8 hours. How long would it take 4 bricklayers to build an identical wall?

2. Leon has 3 dogs, and a bag of dog food lasts 4 days. If he was given another dog to mind, how long would a bag of dog food last?

3. It takes 6 women 6 days to complete a job. How long would it take 2 women to do the same job?

4. If it takes 8 machines 4 days to complete a process, how many machines will it take to complete the process in 12 hours?

5. In a castle, there is enough food for 72 soldiers to last 24 days. After 4 days, another 8 soldiers join them. For how many days will the remaining food last now?

6. CHALLENGE Six men are employed to complete a job in 8 days. After 3 days' work, 2 of the men are taken off the job. Find the total time it takes for the job to be completed from start to finish.

Answers page 126

KEY SKILL

3 Unitary method B

HINTS

- Make sure you know the **Unitary method Hints** from Key Skill 2 on page 12.

Reminder!

- Read the question carefully to identify what needs to be found
- Re-read the question when you've finished your answer to make sure that the question has been answered and your solution makes sense.

Examples

Simon can paint a room in half an hour. It would take Liam twice as long as Simon to paint the same room. How long would it take them to paint the room if they worked together?

Solution

Simon takes 30 minutes. Liam takes 60 minutes.

The LCM of 30 and 60 is 60.

Simon: rooms in 30 minutes = 1

rooms in 60 minutes = 2

Liam: rooms in 60 minutes = 1

Simon and Liam: rooms in 60 minutes = 3

As 3 rooms take 60 minutes, then 1 room would take 60 ÷ 3 = 20 minutes.

∴ they would take 20 minutes working together.

FOCUS on ...

1. The question: Asks you to find the time if they work together.
2. The information: Gives you the time taken for each of them to complete the job.
3. Your working: You need to find a common multiple for both Simon and Liam to find the number of rooms painted in a common time.
   Common multiples of 30 and 60: 60, 120, ...
   Simon: rooms in 30 minutes = 1
   rooms in 60 minutes = 2
   Liam: rooms in 60 minutes = 1
   Then add:
   As 2 + 1 = 3, they complete 3 rooms in 60 minutes. You then divide to find the time for each room. As 60 ÷ 3 = 20, it takes 20 minutes.
4. Your answer: Make sure you write the correct units: It would take 20 minutes.

Josh can weed a garden in 40 minutes. Barb can weed the same garden in 1 hour. How long would it take for them to weed the garden if they worked together but independently?

Solution

LCM of 40 and 60 is 120.

Josh: gardens in 40 minutes = 1

gardens in 2 hours = 3

Barb: gardens in 1 hour = 1

gardens in 2 hours = 2

Josh and Barb: gardens in 2 hours = 5

As 5 gardens take 2 hours, or 120 minutes, then 1 garden would take 120 ÷ 5 = 24 minutes.

∴ they would take 24 minutes working together.

FOCUS on ...

1. The question: Asks you to find the time if they work together.
2. The information: Gives you the time taken for each of them to complete the job.
3. Your working: You need to find a common multiple for both Josh and Barb to find the number of gardens weeded in a common time.
   Common multiples of 40 and 60: 120, 240, ...
   Josh: gardens in 40 minutes = 1
   gardens in 120 minutes = 3
   Barb: gardens in 60 minutes = 1
   gardens in 120 minutes = 2
   Then add:
   As 3 + 2 = 5, they complete 5 gardens in 120 minutes.
   Then divide to find the time for each garden.
   As 120 ÷ 5 = 24, it takes 24 minutes.
4. Your answer: Make sure you write the correct units: It would take 24 minutes.

Now try these!

1. Kirra can paint a fence in 3 hours, while Bradie can do it in 6 hours. How long will it take to paint the fence if they work together?

2. Pipe A can fill a tank in 30 minutes and Pipe B can fill the tank in 50 minutes. How long would it take to fill the tank if both pipes are used together?

3. Mr Palagyi keeps Jason and Seth in for lunchtime detention to clean the desks in his classroom. If Jason cleans by himself he takes 8 minutes. If Seth cleans them by himself he takes 12 minutes. How long will it take them to clean the desks together?

4. An employee can stock a grocery shelf in 15 minutes, while a second employee can stock the same shelf in 12 minutes. How long would it take for both employees to complete the job working together?

5. Lee can groom all the horses in a stable in 4 hours, while Beck completes the same job in one less hour. How long would it take to complete the job together?

6. CHALLENGE It takes Hannah 4 hours to clean her room. When her mother cleans the room it only takes 2 hours. If Hannah started cleaning her room at 3 pm, and her mother started helping her after one hour, what time would they finish?

Answers pages 126–127

KEY SKILL

4 Unitary method C

HINTS

- Make sure you know the **Unitary method Hints** from Key Skill 2 on page 12.

Reminder!

- Read the question carefully to identify what needs to be found
- Re-read the question when you've finished your answer to make sure that the question has been answered and your solution makes sense.

Examples

David drinks a carton of juice in 6 days. David and Sue together drink a carton in 4 days. How long would it take Sue to drink a carton by herself?

Solution

LCM of 6 and 4 is 12.

David: cartons in 12 days = 2

David and Sue: cartons in 12 days = 3

∴ Sue must take 12 days to drink a carton.

FOCUS on...

1. The question: Asks you to find the time it takes Sue to drink a carton by herself.
2. The information: Gives you the time taken for David to drink a carton and for them together to drink a carton.
3. Your working: You need to find a common multiple for David and David and Sue:
   Common multiples of 6 and 4: 12, 24, ...
   David: cartons in 6 days = 1
   cartons in 12 days = 2
   David and Sue: cartons in 4 days = 1
   cartons in 12 days = 3
   Then subtract:
   As 3 − 2 = 1, then David drinks 2 and Sue 1.
4. Your answer: Make sure you write the correct units: Sue takes 12 days to drink a carton.

A father can mow his lawn in 20 minutes. When his son helps him, using another mower, the job can be completed in 8 minutes less time. How long would it take the son if he worked by himself?

Solution

Father: 20 minutes, father and son: 12 minutes

LCM of 20 and 12 is 60.

Father: lawns mowed in 60 minutes = 3

Father and son: lawns mowed in 60 minutes = 5

As 5 − 3 = 2, then the son can mow 2 lawns in 60 minutes, or 1 lawn in 30 minutes.

∴ the son takes 30 minutes to mow the lawn.

FOCUS on...

1. The question: Asks you to find the time it takes for the son to mow the lawn.
2. The information: Gives you the time taken for the father to mow and for both the father and his son to mow a lawn.
3. Your working: You need to find a common multiple for the father and the father and son.
   As 20 − 8 = 12, it takes 12 min when both mowing.
   Common multiples of 20 and 12: 60, 120, ...
   Father: lawns in 20 minutes = 1
   lawns in 60 minutes = 3
   Father and son: lawns in 12 minutes = 1
   lawns in 60 minutes = 5
   Then subtract:
   As 5 − 3 = 2, then the son mows 2 lawns in 60 minutes, or 1 lawn in 30 minutes.
4. Your answer: Make sure you write the correct units: The son takes 30 minutes to mow the lawn.

Now try these!

1. Two men working together can complete a job in 1 hour 20 minutes. The first man working alone can complete the task in 2 hours. How long would it take for the second man to finish the job if he worked alone?

2. When Carmel turns on both the hot and the cold taps it takes 6 minutes to fill the bath. If it takes three times as long to fill the bath with the hot water tap only, how long would it take to fill the bath using the cold water tap alone?

3. Megan takes an hour and a half to rake the lawn. When her sister Hayley helps her it takes only 40 minutes. How long would it take if Hayley raked the lawn by herself?

4. Miles and Millie deliver pamphlets. Miles can deliver a box of pamphlets in half an hour, while Millie and Miles working together take 12 minutes. How long would it take for Milllie if she worked by herself?

5. Andrew and Rose can finish a job together in 3 hours. Andrew can do the job by himself in 5 hours. How long would it take Rose if she worked by herself?

6. CHALLENGE Donald can complete a job in 20 minutes. If he has been working on the job for 10 minutes, then is joined by Daffy and the job is completed in another 6 minutes, how long would it have taken Daffy to do the job by herself?

Answers page 127

KEY SKILL

5 Fractions with unitary method A

HINTS

- The unitary method finds the value of one unit which is then used to get the answer. This 'unit' could be, for example, one-fifth of a quantity,
  e.g. If three-fifths of an object has a mass of 6 kg, what is the mass of one-fifth of the object?
  Mass of one-fifth of the object $= 6 \div 3$
  $= 2$ $\quad \therefore$ 2 kg
  Now, find the mass of the object.
  Mass of five-fifths of the object $= 2 \times 5 = 10$ $\quad \therefore$ 10 kg

Reminder!

- Read the question carefully to identify what needs to be found
- Re-read the question when you've finished your answer to make sure that the question has been answered and your solution makes sense.

Examples

Mitchell originally planned to spend three-quarters of his savings on a shirt. Instead, he spent four-fifths of his savings on a pair of jeans. If the shirt costs \$84, how much was the pair of jeans?

Solution

$$\begin{aligned}\text{Three-quarters of savings} &= 84\\ \text{Four-quarters of savings} &= 84 \div 3 \times 4\\ &= 112\\ \text{Cost of jeans} &= \frac{4}{5} \times 112\\ &= 89.6\end{aligned}$$

$\therefore$ the jeans cost \$89.60.

FOCUS on ...

1. The question: Asks you to find the price of a pair of jeans.
2. The information: Gives you the cost of the shirt that equals three-quarters of Mitchell's savings.
3. Your working: You need to find the total amount of Mitchell's savings:
   Three-quarters of savings = 84
   One-quarter of savings $= 84 \div 3 = 28$
   Four-quarters of savings $= 28 \times 4 = 112$
   Find four-fifths of this amount:
   Cost of jeans $= \frac{4}{5} \times 112 = 89.6$
4. Your answer: Make sure you write the correct units: The pair of jeans cost \$89.60.

A petrol tank is three-fifths full. It takes another 24 litres to completely fill the tank. Later, when the tank is one-quarter full how much petrol is needed to completely fill the tank?

Solution

$$\begin{aligned}\text{Two-fifths of tank} &= 24\\ \text{Five-fifths of tank} &= 24 \div 2 \times 5\\ &= 60\\ \text{Petrol required} &= \frac{3}{4} \times 60\\ &= 45\end{aligned}$$

$\therefore$ 45 litres is needed to fill the tank.

FOCUS on ...

1. The question: Asks you to find the amount of petrol needed to fill the tank
2. The information: Gives you the amount it takes to fill the tank from three-fifths full, and then the amount later remaining in the tank.
3. Your working: You need to find the capacity of the tank:
   As, $1 - \frac{3}{5} = \frac{2}{5}$, two-fifths of tank = 24
   One-fifth of tank $= 24 \div 2 = 12$
   Five-fifths of tank $= 12 \times 5 = 60$
   The capacity of the tank is 60 litres.
   Find three-quarters of the capacity:
   As $1 - \frac{1}{4} = \frac{3}{4}$, Petrol required $= \frac{3}{4} \times 60 = 45$
4. Your answer: Make sure you write the correct units: 45 litres is needed to fill the tank.

Now try these!

1. Each morning James sets out to walk to school but along the way gets a lift with one of his friends. On Monday, he walked 1.2 km which was two-fifths of the distance to school. The following morning he walked two-thirds of the distance to school before he got a lift. How far did he walk on Tuesday morning?

2. Bridget is reading a book. After reading three-eighths of the book she still has 175 pages remaining. How many pages will she have remaining when she has read three-quarters of the book?

3. Each week Phelia spends two-fifths of her weekly allowance on a magazine costing $7.60. If one-tenth of her allowance is saved, how much does she save each week?

4. A water tank is three-twentieths full and currently holds 840 litres. After a heavy storm the water level rises to four-fifths full. How much water is now in the tank?

5. On each morning that Mustaf trains he jogs two-fifths of the total distance, walks one-quarter and runs the remaining distance. If he runs 2.8 km, how far does he walk?

6. CHALLENGE Harley Primary School has 240 students. Two-thirds of the students who attend the school play weekend sport. This number equals two-ninths of the number of students who attend the nearby Harley High School. How many students attend the high school?

Answers pages 127–128

KEY SKILL

6 Fractions with unitary method B

HINTS

- Make sure you know the **Fractions with unitary method Hints** from Key Skill 5 on page 18.

Reminder!

- Read the question carefully to identify what needs to be found
- Re-read the question when you've finished your answer to make sure that the question has been answered and your solution makes sense.

Examples

Lance spent three-quarters of his savings on a holiday and two-thirds of his remaining money on a set of headphones. If he spent $80 on the headphones how much savings did Lance originally have?

Solution

$$\begin{aligned}\text{Fraction for headphones} &= \frac{2}{3} \times (1 - \frac{3}{4}) \\ &= \frac{1}{6} \\ \text{One-sixth of savings} &= 80 \\ \text{Six-sixths of savings} &= 80 \times 6 \\ &= 480\end{aligned}$$

∴ Lance originally saved $480.

FOCUS on ...

1. The question: Asks you to find the original amount of savings.
2. The information: Gives you the fraction spent on a holiday and the fraction of the remainder that equals $80.
3. Your working: You need to find the fraction of savings that equals $80, and then find the whole amount of savings:

$$\begin{aligned}\text{Fraction on holiday} &= \frac{3}{4} \\ \text{Fraction remaining after holiday} &= \frac{1}{4} \\ \text{Fraction on headphones} &= \frac{2}{3} \times \frac{1}{4} = \frac{1}{6} \\ \therefore \text{one-sixth of savings} &= 80 \\ \text{six-sixths of savings} &= 80 \times 6 = 480\end{aligned}$$

4. Your answer: Make sure you write the correct units: Lance originally saved $480.

Telia is travelling from her home to visit her grandmother who lives interstate. On the first day Telia travels three-eighths of the total distance and on the second day she travels three-fifths of the remaining distance. If on the second day she travels 450 km, how far is the entire trip?

Solution

$$\begin{aligned}\text{Fraction of trip on second day} &= \frac{3}{5} \times (1 - \frac{3}{8}) \\ &= \frac{3}{8} \\ \text{Three-eighths of total trip} &= 450 \\ \text{Eight-eighths of the total trip} &= 450 \div 3 \times 8 \\ &= 1200\end{aligned}$$

∴ the total distance is 1200 km.

FOCUS on ...

1. The question: Asks you to find the total distance.
2. The information: Gives you the fraction travelled on day 1, then the fraction and distance for day 2.
3. Your working: You need to find the fraction of the trip covered on day 2 which equals 450 km.

   As $1 - \frac{3}{8} = \frac{5}{8}$, then $\frac{5}{8}$ of trip remains after first day.

   As $\frac{3}{5} \times \frac{5}{8} = \frac{3}{8}$, then $\frac{3}{8}$ of distance is 450 km.

   Then use the unitary method to find the total distance:

$$\begin{aligned}\text{Three-eighths of trip} &= 450 \\ \text{One-eighth of trip} &= 450 \div 3 \\ &= 150 \\ \text{Eight-eighths of trip} &= 150 \times 8 \\ &= 1200\end{aligned}$$

4. Your answer: Make sure you write the correct units: The total distance is 1200 km.

Now try these!

1. Jessica spent two-thirds of her birthday money on a watch and three-quarters of her remaining money on a set of earrings. If she spent $40 on the earrings, what was the original amount of her birthday money?

2. Vincent is taking 3 days to travel from Portville to Lonsdale. On the first day he leaves Portville late and only covers one-third of the total distance. On the second day he travels 480 km which is three-quarters of the remaining distance. What is the total distance of the trip?

3. Seth inherited a sum of money. He decided to give three-quarters of the money to World Vision and one-third of the remaining money to Red Cross. If he gave $120 to Red Cross, what was the total amount he received in the inheritance?

4. In a lolly bag there are three colours of jelly beans. One-quarter are red, one-third of the remainder are black and the rest are white. If there are 6 black jelly beans in the bag, what is the total number of jelly beans in the bag?

5. Two-thirds of Brittany's friends on a social media site are schoolfriends and three-fifths of the remainder are relatives. If she has 18 relatives as friends, what is the total number of her friends on the site?

6. CHALLENGE Julia found a bag of money. When it was unclaimed the police told her she could keep it. She decided to give one-quarter of the money to her parents, one-fifth to her sister and one-eleventh of the remainder to her brother. If her brother received $32, how much did she find?

Answers page 128

KEY SKILL

7 Fractions with unitary method C

HINTS

- Make sure you know the **Fractions with unitary method Hints** from Key Skill 5 on page 18.

Reminder!

- Read the question carefully to identify what needs to be found
- Re-read the question when you've finished your answer to make sure that the question has been answered and your solution makes sense.

Examples

A petrol tank is three-quarters full. After 16 litres is used, the tank is half full. How many litres can the tank hold?

Solution

$$\begin{aligned}\text{Fraction used} &= \frac{3}{4} - \frac{1}{2} \\ &= \frac{1}{4} \\ \text{One-quarter of tank} &= 16 \\ \text{Four-quarters of tank} &= 16 \times 4 \\ &= 64\end{aligned}$$

$\therefore$ the petrol tank can hold 64 litres.

FOCUS on...

1. The question: Asks you to find the capacity of the tank.
2. The information: Gives you the amount of petrol equal to the difference of two fractions.
3. Your working: You need to find the difference in the two fractions.
   As $\frac{3}{4} - \frac{1}{2} = \frac{1}{4}$ then $\frac{1}{4}$ of the total capacity is 16 L.
   Use the unitary method to find the total capacity.
   $$\begin{aligned}\text{One-quarter of tank} &= 16 \\ \text{Four-quarters of tank} &= 16 \times 4 \\ &= 64\end{aligned}$$
4. Your answer: Make sure you write the correct units:
   The tank can hold 64 litres.

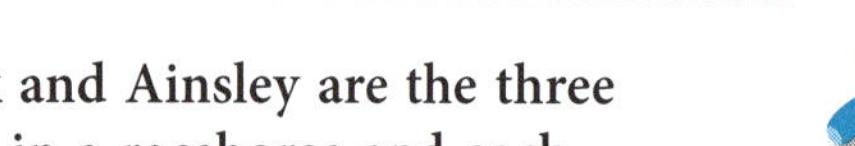

Connor, Jack and Ainsley are the three shareholders in a racehorse and each receive a portion of any prizemoney. From a race win, Connor receives one-fifth of the prizemoney, while Jack receives three-eighths. If Connor's share is \$8400 less than Jack's, how much does Ainsley receive?

Solution

$$\begin{aligned}\text{Difference in Connor and Jack} &= \frac{3}{8} - \frac{1}{5} \\ &= \frac{7}{40}\end{aligned}$$

$$\begin{aligned}\text{Seven-fortieths of money} &= 8400 \\ \text{Forty-fortieths of money} &= 8400 \div 7 \times 40 \\ &= 48\,000\end{aligned}$$

$$\begin{aligned}\text{Ainsley's amount} &= [1 - (\frac{3}{8} + \frac{1}{5})] \times 48\,000 \\ &= 20\,400\end{aligned}$$

$\therefore$ Ainsley receives \$20 400

FOCUS on...

1. The question: Asks you to find Ainsley's share.
2. The information: Gives you the amount of prizemoney equal to the difference of two fractions.
3. Your working: You need to find the difference in the two fractions:
   As $\frac{3}{8} - \frac{1}{5} = \frac{7}{40}$ then $\frac{7}{40}$ of the total is \$8400.
   Use the unitary method to find the total prizemoney:
   $$\begin{aligned}\text{Seven-fortieths of prizemoney} &= 8400 \\ \text{One-fortieth of prizemoney} &= 8400 \div 7 \\ &= 1200 \\ \text{Forty-fortieths of prizemoney} &= 1200 \times 40 \\ &= 48\,000\end{aligned}$$
   Find the fraction that is Ainsley's share:
   $$\begin{aligned}\text{Ainsley's fraction} &= 1 - (\frac{3}{8} + \frac{1}{5}) \\ &= \frac{17}{40}\end{aligned}$$
   Find Ainsley's amount:
   $$\begin{aligned}\text{Ainsley's amount} &= \frac{17}{40} \times 48\,000 \\ &= 20\,400\end{aligned}$$
4. Your answer: Make sure you write the correct units:
   Ainsley receives \$20 400.

Now try these!

1. A soap dispenser is two-thirds full. After 40 mL is used, the dispenser is half full. What amount of soap is in the dispenser when it is full?

2. Sigrid is training for a triathlon and records the distances she runs, swims and cycles. Two-thirds of the total distance is cycling and one-quarter is running. If she cycles 25 km more than she runs, how far does she swim?

3. A wading pool is half filled with water. When 540 litres of water is added, the pool is four-fifths full. How many litres can the pool hold?

4. On the sixth day of her overseas trip Donika calculated she had spent three-tenths of her savings. A week later she only had one-third of her money remaining. If Donika had spent $1155 during that week, what amount of money had she originally saved for the trip?

5. Rowan is travelling between two towns. When he has completed one-quarter of the journey he stops for lunch. After travelling another 280 km he calculates that he has two-fifths of the journey to complete. What is the distance between the two towns?

6. CHALLENGE Students from Years 7, 8 and 9 raised money for a charity. Year 7 students raised one-sixth of the total and Year 9 students raised three-eighths. If Year 8 students contributed $4400, what was raised by Year 9 for the charity?

Answers page 129

REVISION TEST  Level of difficulty—Average

1. In a certain game of cards the losing players need to calculate the total points of the cards remaining in their hand. A heart is worth (+4), diamond (−2), club (−3) and a spade (−4). What is Natasha's total score if she has 3 hearts, a club, 2 spades and 3 diamonds?

2. In 6 days an elephant in a zoo eats 432 kg of food. If the elephant is given the same amount each day how much food is needed for the month of May?

3. It takes 40 minutes for a tanker to be filled with oil when 3 similar hoses are used. How much time would be saved if 2 extra hoses were used?

4. Students in a class baked three varieties of muffins. One-quarter of the muffins were blueberry, one-third choc-chip and the remainder were lemon-poppyseed. If the students made 60 blueberry muffins, how many of the lemon-poppyseed variety were made?

5. A pole is to be painted red, blue and green. Half of the pole is to be painted red, one-third blue and the remaining 1.2 m green. How long is the pole?

6. Fiona is holidaying in Europe. After a week of the trip she calculates she has spent two-fifths of her spending money. During the following week she spends €630 and calculates that she has a quarter of her spending money remaining. What was the original amount of money?

Answers pages 129–130

REVISION TEST  Level of difficulty—Challenging

1. Five workers were employed to complete a job in 6 days. After 2 days' work, 3 of the employees were taken off the job. How many more days will the job take the remaining workers?

2. It takes Mickayla 2 hours to clean her room. When her mother cleans the room it only takes her 30 minutes. If Mickayla started cleaning her room at 4:30 pm, and her mother started helping her at 5:00 pm, what time will they finish?

3. Jack and Jill can paint a fence together in 4 hours. If Jack can do the job by himself in 6 hours, what fraction of the job does Jill do?

4. Michael and Scott are members of a bird club and have the same number of finches. Two-thirds of the 45 birds in Michael's aviary are finches while three-fifths of Scott's birds are finches. What is the total number of birds in Scott's aviary?

5. In Year 8 five-sixths of the girls are right-handed. Two-thirds of the left-handed girls have brown hair. One-fifth of the brown-haired left-handed girls play tennis. If there are 2 brown-haired, left-handed tennis players, how many girls are in Year 8?

6. Isabella started reading a book. On Monday she read three-tenths of the book, on Tuesday, one-eighth, on Wednesday two-fifths and on Thursday one-sixth. If she had read 288 pages on the Wednesday alone, how many pages remain to be read on Friday?

Answers page 130

KEY SKILL

8 Percentages and unitary method

HINTS

- The unitary method finds the value of one unit which is then used to get the answer. This 'unit' could be 1% or 10%, or another percentage, e.g. 60% of a quantity is 42. What is the quantity?

$$\begin{aligned} 60\% \text{ of quantity} &= 42 \\ 10\% \text{ of quantity} &= 42 \div 6 \\ &= 7 \\ 100\% \text{ of quantity} &= 7 \times 10 \\ &= 70 \end{aligned}$$

  ∴ the quantity is 70.

  [You can use a single line: $42 \div 6 \times 10 = 70$; or $42 \div 60 \times 100 = 70$.]

Reminder!

- Read the question carefully to identify what needs to be found.
- Re-read the question when you've finished your answer to make sure that the question has been answered and your solution makes sense.

Examples

Ken spends 23% of his weekly income on his house repayments. If this amounts to $322, find his weekly income.

Solution

$$\begin{aligned} 23\% \text{ of income} &= 322 \\ 100\% \text{ of income} &= 322 \div 23 \times 100 \\ &= 1400 \end{aligned}$$

∴ Ken earns $1400 per week.

FOCUS on ...

1. **The question:** Asks you to find the weekly income.
2. **The information:** Gives you the percentage spent on repayments and what this equals in money.
3. **Your working:** You need to use the unitary method because you need 100% and you know that 23% equals $322.

$$\begin{aligned} 23\% \text{ of income} &= 322 \\ 1\% \text{ of income} &= 322 \div 23 \\ 100\% \text{ of income} &= 322 \div 23 \times 100 \\ &= 1400 \end{aligned}$$

4. **Your answer:** Make sure you write the correct units: Ken earns $1400 per week.

Sam spent 24% of her European holiday in Italy and 16% in Spain. If she spent 12 days in Italy, how much time did she spend in Spain?

Solution

$$\begin{aligned} 24\% \text{ of holiday} &= 12 \\ 16\% \text{ of holiday} &= 12 \div 24 \times 16 \\ &= 8 \end{aligned}$$

∴ Sam spent 8 days in Spain.

FOCUS on ...

1. **The question:** Asks you to find the amount of time spent in Spain.
2. **The information:** Gives you the percentage of time spent in Italy and Spain and the number of days in Italy.
3. **Your working:** You need to use the unitary method because you have 24% and you need 16%.

$$\begin{aligned} 24\% \text{ of holiday} &= 12 \\ 2\% \text{ of holiday} &= 12 \div 12 \\ &= 1 \\ 16\% \text{ of holiday} &= 1 \times 8 \\ &= 8 \end{aligned}$$

4. **Your answer:** Make sure you write the correct units: Sam spent 8 days in Spain.

Now try these!

1. Brayden and some friends went to a restaurant for dinner. At the end of the night Brayden's share of the cost of the meal was \$60 which was 24% of the total bill. What was the total cost of the dinner?

2. A survey of Year 8 electives shows that 45 students chose Food Skills. 20% of the total number of students chose Food Skills and 24% chose Outdoor Recreation. What was the number of students who chose Outdoor Recreation?

3. In his history test Victor scored 68 marks which his teacher converted to 85%. What was the maximum mark for the history test?

4. Liam's doctor told him that he should lose 8% of his body mass which was equal to 6 kg. If he lost 10% of his mass, what is his new mass?

5. An alloy consists of 25% nickel, 40% zinc and copper as the remainder. If a block of the alloy consists of 8 kg of zinc, what is the mass of copper in the block?

6. CHALLENGE The excise tax on petrol was increased from 22.5% to 25% and this brought a 4.5 cents per litre increase in the price. What is the new price of petrol per litre?

Answers pages 130–131

KEY SKILL

9 Increasing and decreasing percentages A

HINTS

- Make sure you know the **Percentages and unitary method Hints** from Key Skill 8 on page 26.
- One whole is 100%.
- An increase is going up,
  e.g. To increase a quantity by 7% means you need to find (100 + 7)%, or 107%, of the quantity. Remember, 107% = 1.07.
- A decrease is going down,
  e.g. To decrease a quantity by 12% means you need to find (100 − 12)%, or 88%, of the quantity. Remember, 88% = 0.88.

Reminder!

- Read the question carefully to identify what needs to be found.
- Re-read the question when you've finished your answer to make sure that the question has been answered and your solution makes sense.

Examples

The price of a house increases by 8%. If the house was valued at $720 000, what is its new value?

Solution

$1.08 \times 720\,000 = 777\,600$

∴ the house is valued at $777 600.

FOCUS on ...

1. The question: Asks you to find the increased price.
2. The information: Gives you the original value and the percentage increase.
3. Your working: You need to increase the original value (100%) by 8%, which means you need to find 108% of the original value.

$$\begin{aligned}\text{New value} &= 108\% \text{ of } 720\,000\\ &= 1.08 \times 720\,000\\ &= 777\,600\end{aligned}$$

4. Your answer: Make sure you write the correct units: The house is now valued at $777 600.

In a discount sale, a shop dropped the price of a laptop by 24%. If the new price was $516.80, what was the previous price?

Solution

$$\begin{aligned}76\% \text{ of previous value} &= 516.80\\ 100\% \text{ of previous value} &= 516.80 \div 76 \times 100\\ &= 680\end{aligned}$$

∴ the laptop was originally priced at $680.

FOCUS on ...

1. The question: Asks you to find the original price before the discount.
2. The information: Gives you the new price and the percentage discount.
3. Your working: You need to use the unitary method because the previous price is 100%, and the discounted price is 24% less than the previous price.

$100 - 24 = 76$

$$\begin{aligned}76\% \text{ of previous value} &= 516.80\\ 1\% \text{ of previous value} &= 516.8 \div 76\\ 100\% \text{ of previous value} &= 516.80 \div 76 \times 100\\ &= 680\end{aligned}$$

4. Your answer: Make sure you write the correct units: The laptop was originally priced at $680.

Now try these!

1. A mass of a box of muesli is normally 760 grams. For a limited time the manufacturer increased the mass of the box and its contents by 20%. What is the mass of the new box of muesli?

2. PetsWorld had a sale on all dog beds. The price of a king-size bed was reduced by 25% to $67.50. What was the original price?

3. The average crowd at the Cats' football ground this season dropped 12% from last year's figure of 14 500. What was the new crowd figure?

4. During March, a store had sales of $806 400. If this was a 12% increase over the February sales, what was the value of the February sales?

5. After getting a discount of 20% Jerome paid $200 for a bicycle. How much did the bicycle originally cost?

6. CHALLENGE A paperback has a recommended retail price (RRP). Eric paid $18 for the book at a store offering a 25% discount off RRP, while Jamie bought it from another store with a 35% discount. What was the difference in price?

Answers page 131

KEY SKILL

10 Increasing and decreasing percentages B

HINTS

- Make sure you know the **Increasing and decreasing percentages Hints** from Key Skill 9 on page 28.
- Successive discounts occur when a shop has already dropped the price of an item by a percentage, and then a further discount occurs,

  e.g. A camera shop had a 20% sale off all lenses, but when a customer offered cash, a further 10% reduction was offered.

Reminder!

- Read the question carefully to identify what needs to be found.
- Re-read the question when you've finished your answer to make sure that the question has been answered and your solution makes sense.

Examples

In March, a manufacturing firm reduced its workforce by 16% from 240 workers. By June a new contract meant that the firm employed new workers and it increased the workforce by 25%. What was the overall change in the workforce?

Solution

$0.84 \times 1.25 \times 240 = 252$

$\therefore$ the workforce increased by 12.

FOCUS on ...

1. The question: Asks you to find the overall change.
2. The information: Gives you the percentage decrease, the percentage increase and the original workforce.
3. Your working: You need to express the percentage decrease and percentage increase as decimals and multiply by the original number.

   $100 - 16 = 84; 100 + 25 = 125$

   $\text{New number} = 0.84 \times 1.25 \times 240 = 252$

   Now subtract to find the difference between the new and the original workforce.

   $252 - 240 = 12$
4. Your answer: The workforce increased by 12.

A hardware store dropped the price of a barbecue by 20%. When the barbecue did not sell, the store discounted the barbecue by another 25%. What was the overall discount, expressed as a percentage?

Solution

$0.80 \times 0.75 = 0.60$

$\therefore$ the new price is 60% of the original price.

$\therefore$ overall discount is 40%.

FOCUS on ...

1. The question: Asks you to find the overall percentage discount.
2. The information: Gives you the two successive percentage discounts.
3. Your working: You need to express the percentage decreases as decimals and multiply.

   $100 - 20 = 80; 100 - 25 = 75$

   $\text{Change} = 0.8 \times 0.75 = 0.6 = 60\%$

   The new price is 60% of the original.

   Subtract this answer from 1 (or 100%).

   As $100 - 60 = 40$, then the new price has decreased by 40%.
4. Your answer: Make sure you write the correct units: The discount is 40%.

Now try these!

1. The original price of a set of headphones was $48 and the storeowner increased the price by 20%, before dropping the price of the headphones by 25%. What was the overall change in the price?

2. Caiphas is overweight and starts a fitness program to lose some weight. In the first month his weight drops by 3% and in the following month it drops by another 2%. What is his overall loss as a percentage?

3. When Brittany was 12 years old her parents gave her $10 per week allowance. On her thirteenth birthday, her allowance increased by 20%, and then when she turned 14 she received a 25% increase. What is her allowance as a 14-year-old?

4. In a slump the population of a mining town dropped by 8% but during the following boom years it increased by 10%. What was the overall percentage change in the population?

5. Three months ago the price of a plane trip from Melbourne to Newcastle was $59. Two months ago the price rose by 10%. Last month the price rose by 5% and yesterday it increased by 2.5%. What is the price of a ticket today?

6. CHALLENGE Lauren is currently paid an allowance of $500 per year. Her father gives her the choice of a 3% allowance rise now or a 1% rise now, a 2% rise in a year's time and then a 3% rise in two years' time. What is the best choice to earn her the most money after 3 full years?

Answers page 131

KEY SKILL

11 Financial mathematics: Profit and loss A

HINTS

- A profit, or a loss, is the difference between the cost price and the selling price.
  A profit is made when selling price > cost price.
  A loss is made when selling price < cost price.
- A profit is made when the cost price is increased to give the selling price.
  A loss is made when the cost price is reduced to give the selling price.
- To write a fraction as a percentage you multiply by 100%.

Reminder!

- Read the question carefully to identify what needs to be found.
- Re-read the question when you've finished your answer to make sure that the question has been answered and your solution makes sense.

Examples

At a garage sale, an antique dealer purchases a chair for $40 and later sells it for $65. What is the profit as a percentage of the cost price when the chair is sold?

Solution

$$\begin{aligned}\text{Profit} &= 65 - 40\\ &= 25\\ \text{Profit \%} &= \frac{25}{40} \times 100\%\\ &= 62.5\end{aligned}$$

$\therefore$ the dealer makes a profit of 62.5%.

FOCUS on ...

1. The question: Asks you to find the profit made.
2. The information: Gives you the cost price and the selling price of the chair.
3. Your working: You need to find the profit.
   $$\begin{aligned}\text{Profit} &= \text{Selling Price} - \text{Cost Price}\\ &= 65 - 40\\ &= 25\end{aligned}$$
   Write this profit as a percentage of the cost price.
   $$\begin{aligned}\text{Profit \%} &= \frac{25}{40} \times 100\%\\ &= 62.5\end{aligned}$$
4. Your answer: Make sure you write the correct units: The dealer makes a profit of 62.5%.

A car was originally purchased for $37 500. After 5 years the owner sold it at a loss of 72% on his purchase price. What was the selling price of the car?

Solution

$$\begin{aligned}\text{Loss} &= 72\% \text{ of } 37\,500\\ &= 0.72 \times 37\,500\\ &= 27\,000\end{aligned}$$

$\therefore$ the loss was $27 000.

$$\begin{aligned}\text{Selling price} &= \text{Cost price} - \text{Loss}\\ &= 37\,500 - 27\,000\\ &= 10\,500\end{aligned}$$

$\therefore$ the car was sold for $10 500.

FOCUS on ...

1. The question: Asks you to find the selling price.
2. The information: Gives you the cost price and the loss as a percentage.
3. Your working: You need to find the amount of loss by finding the percentage of the cost price.
   $$\begin{aligned}\text{Loss} &= 72\% \text{ of } 37\,500\\ &= 0.72 \times 37\,500\\ &= 27\,000\end{aligned}$$
   $\therefore$ the loss was $27 000.
   The loss is subtracted from the cost price to find the selling price.
   $$\begin{aligned}\text{Selling price} &= \text{Cost price} - \text{Loss}\\ &= 37\,500 - 27\,000\\ &= 10\,500\end{aligned}$$
4. Your answer: Make sure you write the correct units: The car was sold for $10 500.

Now try these!

1. A painting is purchased for \$120 000 and 4 years later is sold for \$186 000. What is the profit as a percentage of the cost price?

2. A family buys an eight-week-old german shepherd pup for \$800. After 2 years the family move into rental accommodation and have to sell the dog for a loss of 45%. At what price do the family sell the dog?

3. Carl buys a motor bike for \$660. He spends \$240 on repairs and sells it for \$1200. What is his percentage profit?

4. Don purchased a box of apples. He bought them at the rate of 5 apples for \$4 and sold them at 4 apples for \$5. What was his profit expressed as a percentage of the cost price?

5. Mitch bought and sold a camera lens. The ratio of his cost price to selling price was 4 : 5. Later he sold a camera tripod for the same percentage profit. If the tripod had originally cost \$120, what was the selling price of the tripod?

6. CHALLENGE The manager of Nikki's Nuts bought some nuts from a producer. He purchased 30 kg of nuts at \$4 per kilogram and another 20 kg of nuts at \$5 per kilogram. If the nuts are to be mixed and sold to make a 75% profit, what will be the sale price per kilogram?

Answers page 132

KEY SKILL

12 Financial mathematics: Profit and loss B

HINTS

- Make sure you know the **Financial mathematics: Profit and loss Hints** from Key Skill 11 on page 32.
- Make sure you know the **Percentages and unitary method Hints** from Key Skill 8 on page 26.

Reminder!

- Read the question carefully to identify what needs to be found.
- Re-read the question when you've finished your answer to make sure that the question has been answered and your solution makes sense.

Examples

By selling a vacuum cleaner for \$480, a store makes a 20% profit on the cost price. At what price should the cleaner be sold to make a profit of 30%?

Solution

120% of cost price = 480

130% of cost price = 480 ÷ 120 × 130

= 520

∴ the vacuum cleaner should be sold for \$520.

FOCUS on ...

1. **The question:** Asks you to find the price to be charged to make a 30% profit.
2. **The information:** Gives you the price of the vacuum cleaner to make a 20% profit.
3. **Your working:** You need to use the unitary method because the previous price is 20% above the cost price, and you need to find 30% above the cost price.
   100 + 20 = 120; 100 + 30 = 130
   120% of cost price = 480
   1% of cost price = 480 ÷ 120
   130% of cost price = 480 ÷ 120 × 130
   = 520
4. **Your answer:** Make sure you write the correct units: The vacuum cleaner needs to be priced at \$520.

If a shopowner had sold a television for \$448 she would have made a 40% profit. Instead she sold the television at a 15% loss. At what price was the television sold?

Solution

140% of cost price = 448

85% of cost price = 448 ÷ 140 × 85

= 272

∴ the television was sold for \$272.

FOCUS on ...

1. **The question:** Asks you to find the selling price at a 15% loss.
2. **The information:** Gives you the selling price at a 40% profit.
3. **Your working:** You need to use the unitary method because the previous price is 40% above the cost price, and you need to find 15% below the cost price.
   100 + 40 = 140; 100 − 15 = 85
   140% of cost price = 448
   1% of cost price = 448 ÷ 140
   85% of cost price = 448 ÷ 140 × 85
   = 272
4. **Your answer:** Make sure you write the correct units: The television was sold for \$272.

Now try these!

1. When a shopowner sells a pair of shoes for \$80 she makes a profit of 25%. At what price should the shoes be sold to make a profit of 40%?

2. The price of Christmas decorations included a profit of 40% on cost price. On Boxing Day the prices are slashed so the store sells items at '10% below cost'. If a tree was originally priced at \$56, what is its Boxing Day price?

3. In May, an electrical shop priced a toaster at \$24 which would make a profit of 20% on the cost price. What will be its new price in June if the store holds a '5% on cost' sale?

4. When a bag is priced at \$75 the stallholder makes a profit of 80%. What will be the price of the bag when the stallholder makes a profit of 20%?

5. The owner of a stationery shop has a 60% profit margin on all the items in her shop. She has a 'Cost + 10% Easter sale'. When a calligraphy set still does not sell she reduces the price by a further 25%. If the original price of the calligraphy set was \$80, what is the final price?

6. CHALLENGE All vehicles at Sam's Quality Cars are priced at a profit of 40% on cost price. When Tony bought a car from the dealership he bargained the price down and received a discount of 20% on the sale price. If the dealership still made a profit of \$3600 on the sale, what was the original cost of the car for the dealership?

Answers page 132

KEY SKILL

13 Financial mathematics: GST

HINTS

- The Goods and Services Tax (GST) is currently set at 10% and so increases the price of goods and services by 10%,
  e.g. Increase $150 by 10%.
  As 100 + 10 = 110, then
  110% of 150 = 1.1 × 150
  = 165
  The new price is $165.
- To find the amount of GST included in a price you divide by 11,
  e.g. How much GST is in a price of a $165 watch?
  165 ÷ 11 = 15
  The GST is $15.

Reminder!

- Read the question carefully to identify what needs to be found.
- Re-read the question when you've finished your answer to make sure that the question has been answered and your solution makes sense.

Examples

A mechanic works for 6 hours to fix a car engine. He charges $70 per hour plus 10% GST. What is the total cost including GST?

Solution

100 + 10 = 110
Hourly rate incl. GST = 1.1 × 70
= 77
Total cost = 77 × 6
= 462
∴ the total cost is $462.

FOCUS on…

1. The question: Asks you to find the total cost including the GST.
2. The information: Gives you the hourly rate, the percentage GST and the number of hours.
3. Your working: You need to find the mechanic's hourly rate including GST.
   100 + 10 = 110
   Hourly rate incl. GST = 110% of 70
   = 1.1 × 70
   = 77
   Then multiply that amount by the number of hours worked.
   Total cost = 77 × 6
   = 462
4. Your answer: Make sure you write the correct units:
   The total cost is $462.

An air-conditioner is priced at $792 which includes a GST of 10%. If the government increases the GST by 2.5%, what is the new price of the air-conditioner?

Solution

100 + 10 = 110; 100 + 12.5 = 112.5
110% of cost price = 792
112.5% of cost price = 792 ÷ 110 × 112.5
= 810
∴ the new price is $810.

FOCUS on…

1. The question: Asks you to find the new price after an increase in GST.
2. The information: Gives you the price including a 10% GST and the amount of increase in GST.
3. Your working: You need to use the unitary method because the previous price is 10% above the non-GST price, and you need to find 12.5% above the non-GST price.
   100 + 10 = 110; 100 + 12.5 = 112.5
   110% of cost price = 792
   1% of cost price = 792 ÷ 110
   112.5% of cost price = 792 ÷ 110 × 112.5
   = 810
4. Your answer: Make sure you write the correct units:
   The new price is $810.

Now try these!

1. A plumber charges \$80 per hour plus GST. How much does he charge for working for 3 hours?

2. If GST is 10%, the price of a ream of paper is \$5.50. If the GST is increased to 15%, find the new price.

3. Mia employs a repairman for some work in her home office. He charges \$65 per hour plus GST and works for 4 hours. To complete the work he buys \$132 worth of hardware products, which includes GST. What is the total amount of GST Mia has paid for the job to be completed?

4. Jo increases the price of a television set by 20% to \$840. What has been the increase in the amount of GST in the price?

5. If the GST increases from 10% to 12% the price of a textbook rises by \$1.32. What is the original price?

6. CHALLENGE Lana's Jewellery store is having a sale. The prices of all diamond rings drop by 20%. On a particular engagement ring the amount of GST drops by \$70.40. What is the discounted price of the ring?

Answers pages 132–133

REVISION TEST 3 Level of difficulty—Average

1. At the gym, Sarah spent 40% of the time on the treadmill, 35% on the weights and the remainder in the pool. If she swam for 15 minutes, how long was Sarah on the treadmill?

2. At a '15%-off end-of-season sale', Matt bought a pair of footy boots. If he paid $102 for the boots, what was the original price?

3. A block of land in 2005 was valued at $190 000. By 2010 the value had dropped by 4.5%, but 5 years later had risen in price by 25%. What was the value of the land in 2015?

4. On his 40th birthday Tom purchased a motor bike for $25 000. After Tom injured his leg in an accident he sold the bike at a loss of 35%. At what price did Tom sell his bike?

5. Everything in Norm's electrical store is priced at 40% above cost price. A television is priced at $840. If the store discounts prices by 25%, what profit will be made on the sale of the television?

6. A book originally priced at $12 is discounted by 20%. If GST is set at 10%, what is the reduction in the amount of GST?

Answers page 133

REVISION TEST  Level of difficulty—Challenging

1. A jewellery store changed its storewide discount sale from '10% off' to '12.5% off'. The price of an engagement ring dropped by \$35. What was the new price of the ring?

2. The manufacturer of a tablet computer recommends the price of \$420. The tablet is on sale for a discount of 15% at Laddies 'Lectrical while at Ed's Discount the price has been reduced to \$336. What is the difference in the two prices?

3. A barbecue was reduced in price by 25%. The store manager then dropped the price by another 20%. What is the comparable single discount offered on the original price?

4. The Sweet Things manager purchased 25 kg of chocolate bullets for \$5.60/kg and 45 kg of chocolate buttons for \$4.90/kg. She then combined the chocolates into a mixture. What price will she sell the mixture for if she wants to make a 60% profit?

5. Sasha bought a camera online and later sold it for \$1392 which gave her a profit of 16% on her cost price. What was the profit expressed as a percentage of the selling price?

6. The Goods and Services Tax (GST) is set at 10%. The price of a refrigerator in an electrical store is reduced by 15%. If the amount of GST on the new price is \$85, what was the original price of the refrigerator?

Answers pages 133–134

KEY SKILL

14 Ratio and rates: Simplifying ratios A

HINTS

- A ratio is the comparison of two quantities with the same units.
- Ratios are written as $a:b$, but could also be expressed as $\frac{a}{b}$.
- A proportion is a statement of equality between two ratios, such as $a:b = c:d$, or $\frac{a}{b} = \frac{c}{d}$.
- Ratios are simplified by dividing (or multiplying) the components of the ratio to form an equivalent ratio, e.g. Simplify $4:18 = 2:9$.

Reminder!

- Read the question carefully to identify what needs to be found.
- Re-read the question when you've finished your answer to make sure that the question has been answered and your solution makes sense.

Examples

In an election, 35% of people voted for White, 25% voted for Black and the remainder voted for Brown. What was the ratio of Brown votes to White votes?

Solution

As $100 - (35 + 25) = 40$, then 40% voted for Brown.

$$\begin{aligned}\text{Ratio} &= 40:35\\ &= 8:7\end{aligned}$$

$\therefore$ the ratio was $8:7$.

FOCUS on ...

1. **The question:** Asks you to find the ratio of Brown votes to White votes.
2. **The information:** Gives you the percentages of votes for White and Black.
3. **Your working:** You need to subtract the total of the percentages for White and Black from 100 to find the percentage who voted for Brown.

   $$\begin{aligned}\text{\% voted for Brown} &= 100 - (35 + 25)\\ &= 100 - 60\\ &= 40\end{aligned}$$

   $\therefore$ 40% voted for Brown.

   The ratio of percentages for Brown to White can be simplified.

   $$\begin{aligned}\therefore \text{Brown} : \text{White} &= 40:35\\ &= 8:7\end{aligned}$$

4. **Your answer:** The ratio was $8:7$.

Martin was four-and-a-half years old when his sister Rose was born. Rose celebrates her sixth birthday today. What is the ratio of Martin's age to Rose's age today?

Solution

As $4\frac{1}{2} + 6 = 10\frac{1}{2}$, Martin is $10\frac{1}{2}$ years old today.

$$\begin{aligned}\text{Ratio} &= 10\tfrac{1}{2}:6\\ &= 21:12\\ &= 7:4\end{aligned}$$

$\therefore$ the ratio is $7:4$.

FOCUS on ...

1. **The question:** Asks you to find the ratio of Martin's age to Rose's age.
2. **The information:** Gives you the age of Rose and the difference in the ages of Martin and Rose.
3. **Your working:** You need to find the age of Martin today.

   Martin's age $= 4\frac{1}{2} + 6 = 10\frac{1}{2}$

   $\therefore$ Martin is $10\frac{1}{2}$ years old today.

   Then simplify the two ages.

   $$\begin{aligned}\text{Ratio of ages} &= 10\tfrac{1}{2}:6\\ &= \frac{21}{2}:\frac{12}{2}\\ &= 21:12\\ &= 7:4\end{aligned}$$

4. **Your answer:** The ratio is $7:4$.

Now try these!

1. A golf club has 1200 members and 950 of them are financial. What is the ratio of financial to unfinancial members?

2. A baker makes 8 dozen Easter buns. Four-and-a-half dozen are made with fruit and the remainder are made with choc-chips. What is the ratio of fruit buns to choc-chip buns?

3. A caravan is priced at $64 000. It is reduced in price by $8000. Find the ratio of the old price to the new price.

4. During her 3-km exercise this morning, Nicole walked for 500 m, ran for 1.2 km and jogged the remainder. What was the ratio of jogging to running?

5. Timothy is reading a 400-page book. He has already read 42% of the book. If this weekend he reads 20% of the total book, what will be the ratio of unread to read pages on Monday?

6. CHALLENGE There are 28 students on the 8M1 class roll. On a particular day a quarter of the students are on an excursion. One student is absent and 15% of the remainder are helping the principal. What is the ratio of those students present in class to those who are on the excursion?

Answers page 134

KEY SKILL

15 Ratio and rates: Simplifying ratios B

HINTS

- Make sure you know the **Ratio and rates: Simplifying ratios Hints** from Key Skill 14 on page 40.
- To combine two ratios together you equate one component in each ratio,
  e.g. A bag contains red, blue and green balls, where the ratio of red to blue is 2 : 3 and blue to green is 3 : 4. What is the ratio of red : blue : green?
  red : blue : green = 2 : 3 : 4

Reminder!

- Read the question carefully to identify what needs to be found.
- Re-read the question when you've finished your answer to make sure that the question has been answered and your solution makes sense.

Examples

In a nut mixture, the ratio of peanuts to cashews is 4 : 3 and the ratio of cashews to almonds is 5 : 6. What is the ratio of peanuts to cashews to almonds?

Solution

peanuts : cashews = 4 : 3
= 20 : 15 (by multiplying by 5)
cashews : almonds = 5 : 6
= 15 : 18 (by multiplying by 3)
∴ peanuts : cashews : almonds = 20 : 15 : 18

FOCUS on ...

1. The question: Asks you to find the ratio of peanuts to cashews to almonds.
2. The information: Gives you the ratio of peanuts to cashews and the ratio of cashews to almonds.
3. Your working: You need to identify the common term—'cashews' is mentioned in both ratios. You multiply both ratios so as to have the same number representing cashews in both ratios. The ratio can then be expressed:
   peanuts : cashews = 4 : 3
   = 20 : 15 (multiply by 5)
   cashews : almonds = 5 : 6
   = 15 : 18 (multiply by 3)
4. Your answer: The ratio is 20 : 15 : 18.

Chrissy recorded the birdlife on a pond in the local park. The ratio of ducks to swans was 7 : 2 and the ratio of swans to egrets was 3 : 4. What was the ratio of ducks to egrets?

Solution

ducks : swans = 7 : 2
= 21 : 6 (by multiplying by 3)
swans : egrets = 3 : 4
= 6 : 8 (by multiplying by 2)
∴ ducks : swans : egrets = 21 : 6 : 8
∴ ducks : egrets = 21 : 8

FOCUS on ...

1. The question: Asks you to find the ratio of ducks to egrets.
2. The information: Gives you the ratio of ducks to swans and the ratio of swans to egrets.
3. Your working: You need to identify the common term—'swans' is mentioned in both ratios. You multiply both ratios so as to have the same number representing swans in both ratios. The ratio can then be expressed:
   ducks : swans = 7 : 2
   = 21 : 6 (multiply by 3)
   swans : egrets = 3 : 4
   = 6 : 8 (multiply by 2)
   ∴ ducks : swans : egrets = 21 : 6 : 8
   ∴ ducks : egrets = 21 : 8
4. Your answer: The ratio was 21 : 8.

Now try these!

1. In a bowl of jelly beans, the ratio of red to blue jelly beans is 3 : 2 and the ratio of blue to black is 4 : 5. What is the ratio of red to blue to black jelly beans?

2. A coffee shop recorded the number of beverages sold one morning. The ratio of coffees to iced chocolates was 15 : 2 and the ratio of iced chocolates to teas was 3 : 10. What was the ratio of coffees to teas?

3. The manager of a newsagency compared her stock of pens. The ratio of red pens to blue pens was 3 : 2 and the ratio of blue pens to black pens was 3 : 5. What was the ratio of red pens to blue pens to black pens?

4. In our soccer games this season, the ratio of wins to draws is 3 : 2 and the ratio of wins to losses is 2 : 1. What is the ratio of draws to losses?

5. Theo, Ronni and Ted collect stamps. The ratio of Theo's stamps to Ronni's stamps is 5 : 3. The ratio of Ronni's stamps to Ted's stamps is 6 : 5. If Theo has 600 stamps, how many stamps has Ted?

6. CHALLENGE A bag contains balls of four different colours. The ratio of green to red is 2 : 1, the ratio of red to blue is 3 : 4 and the ratio of blue to yellow is 2 : 5. Explain why the ratio of green to yellow balls is 3 : 5.

Answers pages 134–135

KEY SKILL

16 Ratio and rates: Dividing quantities in a given ratio

HINTS

- Ratios are simplified by dividing (or multiplying) the components of the ratio to form an equivalent ratio, e.g. Simplify $4:18 = 2:9$.
- The components of a ratio can be referred to as parts, e.g. In a group, the ratio of boys : girls is $3:5$. This means there is a total of 8 parts: $\frac{3}{8}$ are boys and $\frac{5}{8}$ are girls.

Reminder!

- Read the question carefully to identify what needs to be found.
- Re-read the question when you've finished your answer to make sure that the question has been answered and your solution makes sense.

Examples

Chloe is 9 years old and her brother Lucas is 12 years old. The pocket money they receive is in the ratio of their ages. Altogether they receive \$28 per week. How much does Chloe receive?

Solution

$$\begin{aligned}\text{Ratio of ages} &= 9:12\\ &= 3:4\\ \text{Total parts} &= 3 + 4\\ &= 7\\ \text{Chloe's amount} &= \frac{3}{7} \times 28\\ &= 12\end{aligned}$$

$\therefore$ Chloe's pocket money is \$12.

FOCUS on ...

1. **The question:** Asks you to find the amount of Chloe's pocket money.
2. **The information:** Gives you the total pocket money and their ages.
3. **Your working:** You need to simplify the ratio of their ages: $12:9 = 4:3$.
   $\therefore$ ratio of pocket money of Chloe : Lucas is $3:4$.
   Find the total number of parts in the ratio.
   Total parts $= 3 + 4 = 7$
   The order is important in ratio questions: Chloe and 3 are both mentioned first. This means 3 out of every 7 dollars of total pocket money goes to Chloe.
   $$\begin{aligned}\text{Chloe's share} &= \frac{3}{7} \times 28\\ &= 12\end{aligned}$$
4. **Your answer:** Make sure you write the correct units: Chloe is given \$12 pocket money.

James has collected 36 stamps. He wants to give away the stamps to 3 friends in the ratio of $2:3:4$. How many stamps does he give to each friend?

Solution

$$\begin{aligned}\text{Total parts} &= 2 + 3 + 4\\ &= 9\end{aligned}$$

$\frac{2}{9} \times 36 = 8$

$\frac{3}{9} \times 36 = 12$

$\frac{4}{9} \times 36 = 16$

$\therefore$ he gives his three friends 8, 12 and 16 stamps.

FOCUS on ...

1. **The question:** Asks you to find the number of stamps given to 3 friends.
2. **The information:** Gives you the total number of stamps and the ratio of the stamps of 3 friends.
3. **Your working:** You need to find the total number of parts in the ratio.
   Total parts $= 2 + 3 + 4 = 9$
   Then find the fractions of the total given to each friend.
   2 parts $= \frac{2}{9} \times 36 = 8$
   3 parts $= \frac{3}{9} \times 36 = 12$
   4 parts $= \frac{4}{9} \times 36 = 16$
4. **Your answer:** Make sure you write the correct units: The friends receive 8, 12 and 16 stamps.

Now try these!

1. Mia and Sienna together bought a $10 lottery ticket. Mia contributed $6 and Sienna the remainder. If they share a win of $25 000 in the ratio of their contributions, how much will Sienna receive?

2. A brand of fertiliser is mixed from amounts of nitrates, potash and phosphates in the ratio of 2 : 3 : 4. How many kilograms of each are in a 45-kilogram bag?

3. Two women begin a business by investing $30 000 and $42 000 respectively. Profits from the business are divided in the ratio of their investments. If the first month's profit totals $8400, what amount does each woman receive?

4. The lengths of a triangle are in the ratio 3 : 5 : 4. Find the lengths of each side if the perimeter of the triangle is 36 cm.

5. A rectangle has a perimeter of 56 cm. If the ratio of its length and width is 4 : 3, find its area.

6. CHALLENGE The ratio of the populations of Town A and Town B is 5 : 3 and the ratio of the populations of Town B and Town C is 4 : 7. If the total population of the three towns is 15 900, find the population of each town.

Answers page 135

KEY SKILL

17 Ratio and rates: Ratio involving unitary method

HINTS

- Make sure you know the **Unitary method Hints** from Key Skill 2 on page 12 and the **Ratio and rates: Simplifying ratios Hints** from Key Skill 14 on page 40.

Reminder!

- Read the question carefully to identify what needs to be found.
- Re-read the question when you've finished your answer to make sure that the question has been answered and your solution makes sense.

Examples

A coffee blend is made from two types of beans from Brazil and Colombia in the ratio 3 : 2. If a bag contains 450 grams of Brazilian coffee, what is the mass of Colombian coffee in the bag?

Solution

$$\begin{aligned} 3 \text{ parts} &= 450 \\ 1 \text{ part} &= 150 \\ 2 \text{ parts} &= 300 \end{aligned}$$

∴ the mixture has 300 grams of Colombian coffee.

FOCUS on ...

1. The question: Asks you to find the mass of Colombian coffee in the blend.
2. The information: Gives you the ratio of Brazilian coffee to Columbian coffee and the mass of Brazilian coffee in the blend.
3. Your working: The order is important in ratio questions. Brazil and 3 are both mentioned first. You use the unitary method and start with what is known.
   $$\begin{aligned} 3 \text{ parts} &= 450 \\ 1 \text{ part} &= 150 \\ 2 \text{ parts} &= 150 \times 2 \\ &= 300 \end{aligned}$$
4. Your answer: Make sure you write the correct units: The mixture has 300 grams of Colombian coffee.

A metal alloy was made by mixing tin, copper and zinc in the ratio 4 : 5 : 3. If 60 kg of copper was used, how much tin and zinc were used?

Solution

$$\begin{aligned} 5 \text{ parts} &= 60 \\ 1 \text{ part} &= 12 \\ 4 \text{ parts} &= 48 \\ 3 \text{ parts} &= 36 \end{aligned}$$

∴ the alloy contains 48 kg of tin and 36 kg of zinc.

FOCUS on ...

1. The question: Asks you to find the mass of tin and zinc in the alloy.
2. The information: Gives you the ratio of tin to copper to zinc and the mass of copper.
3. Your working: The order is important in ratio questions. Copper and 5 are both mentioned second. You use the unitary method and start with what is known.
   $$\begin{aligned} 5 \text{ parts} &= 60 \\ 1 \text{ part} &= 12 \\ 4 \text{ parts} &= 12 \times 4 \\ &= 48 \\ 3 \text{ parts} &= 12 \times 3 \\ &= 36 \end{aligned}$$
4. Your answer: Make sure you write the correct units: The alloy contains 48 kg of tin and 36 kg of zinc.

Now try these!

1. Two sides of a rectangle are in the ratio 3 : 5. If the longer side is 60 cm, what is the length of the shorter side of the rectangle?

2. Concrete is made by mixing gravel, sand and cement in the ratio 5 : 3 : 2. How much sand and cement would be required for a mixture with 400 kg of gravel?

3. A sum of money was divided between three sisters in the ratio 8 : 5 : 3. If the smallest share was $660, what were the other shares and the total amount of money?

4. The sides of a scalene triangle are in the ratio of 3 : 4 : 5. If the smallest side is 8.1 cm, what is the length of the longest side?

5. When Michelle sold a necklace the ratio of cost price to selling price was 4 : 9. If she made a profit of $65, what were her cost and selling prices?

6. CHALLENGE In her science lesson, Mrs Bradbury made a chemical solution by mixing an acid with water in the ratio of 1 : 4. Jack took the solution and mixed it with water in the ratio of 1 : 3. If Jack added 120 mL of water to make the final solution, how much acid did Mrs Bradbury use?

Answers pages 135–136

KEY SKILL

18 Ratio and rates: Map scale

HINTS

- Map scale refers to the ratio of the distance on a map to the corresponding distance on the ground,
  e.g. A commonly used ratio is the scale of 1 : 100 000.
  This means 1 cm = 100 000 cm
  ∴ 1 cm = 1000 m
  ∴ 1 cm = 1 km

Reminder!

- Read the question carefully to identify what needs to be found.
- Re-read the question when you've finished your answer to make sure that the question has been answered and your solution makes sense.

Examples

A map is drawn to a scale of 1 : 100 000. Two petrol stations are separated by a straight road. The two petrol stations are located 26 km apart. Find the distance, in centimetres, between the two petrol stations on the map.

Solution

Distance in cm = 26 × 1000 × 100
= 2 600 000

∴ the two petrol stations are 2 600 000 cm apart.

Distance on map = 2 600 000 ÷ 100 000
= 26

∴ the two petrol stations are 26 cm apart on the map.

FOCUS on ...

1. **The question:** Asks you to find the distance between two petrol stations on a map.
2. **The information:** Gives you the scale of the map and the distance between the two petrol stations on the ground.
3. **Your working:** Change the distance from kilometres to centimetres.
   Distance in cm = 26 × 1000 × 100
   = 2 600 000
   ∴ the two petrol stations are 2 600 000 cm apart.
   Use the scale to divide the ground distance by 100 000.
   Distance on map = 2 600 000 ÷ 100 000
   = 26
4. **Your answer:** Make sure you write the correct units:
   The two petrol stations are 26 cm apart on the map.

The scale on a map is 1 : 50 000. The distance from Carmel Corner to Jacks Flat is measured as 13 cm. How far apart are the two locations in kilometres?

Solution

Distance = 13 × 50 000
= 650 000

∴ the two locations are 650 000 cm apart.

As 650 000 ÷ 100 = 6500 and
6500 ÷ 1000 = 6.5,
the two locations are 6.5 km apart.

FOCUS on ...

1. **The question:** Asks you to find the distance between two locations on the ground.
2. **The information:** Gives you the scale of the map and the distance between the two locations on the map.
3. **Your working:** Use the scale to multiply the map distance by 50 000.
   Distance = 13 × 50 000
   = 650 000
   ∴ the two locations are 650 000 cm apart.
   Divide by 100 000 to change the centimetres to kilometres.
   As 650 000 ÷ 100 000 = 6.5, the two locations are 6.5 km apart.
4. **Your answer:** Make sure you write the correct units:
   The two locations are 6.5 km apart.

Now try these!

1. A scale of 1 : 1 000 000 is used to draw a map of a country. If the distance from Town A to Town B is 48 km, how far apart are the towns on the map, in centimetres?

2. Two lookouts are 26 cm apart on a map. The map has a scale of 1 : 250 000. Find the distance, in kilometres, between the two lookouts.

3. Two towns are 120 km apart and are shown on a map as being 30 cm apart. Find the scale used on the map and the actual distance between two villages which are 18 cm apart on the map.

4. The scale used on a map is 1: 25 000. The dimensions of a parcel of land on the map are 8.5 cm by 6.4 cm. Find in square kilometres the actual area of the land.

5. Two maps are drawn for the same area of land. The scale on Map A is 1 : 8000 while the scale on Map B is 1 : 12 000. Two towns are 24 cm apart on Map A. How far are they apart on Map B?

6. CHALLENGE A map has been drawn to the scale of 1: 1000. A park is shown on the map with an area of 250 cm^2. What is the actual area of the park on the ground, in hectares?

Answers page 136

REVISION TEST 5 Level of difficulty—Average

1. One angle in a triangle is 60°. If the other two angles are in the ratio of 3 : 2, what is the size of the largest angle?

2. Up until today, Noah has read a third of the pages in his book. If today he reads half the total pages in the book, what is the ratio of pages read to unread?

3. A breakfast cereal contains rice, wheat and oats in the ratio 2 : 4 : 5. If the mixture contains 120 grams of wheat, what is the mass of the rice and oats in the cereal?

4. The ratio of Tom's age to Mia's age is 4 : 3 and the ratio of Mia's age to Aiden's age is 2 : 5. If Aiden is 30 years old, how old will Tom be in 4 years?

5. Two towers are 17 cm apart on a map. If the scale used on the map is 1 : 25 000, how far apart are the towers?

6. Ben is 20% taller than Jack and his brother Ethan is 40% taller than Jack. Write the ratio of the heights of Ben to Jack to Ethan.

Answers page 137

REVISION TEST  Level of difficulty—Challenging

1. The ratio of the perimeter to the length of a rectangle is 8 : 3. If the perimeter of the rectangle is 120 cm, what is the area of the rectangle?

2. The angles in a triangle are in the ratio of 4 : 3 : 2. What is the difference between the smallest and the largest angle?

3. A father distributes a sum of money to Len, Ken, Jen and Ben in the ratio 2 : 3 : 4 : 5 respectively. If Ken gets $150 less than Ben, how much money do Len and Jen receive?

4. Three friends were comparing their results on a history test. The ratio of Sophia's mark to Ava's mark was 12 : 7 and the ratio of Sophia's mark to Ella's mark was 3 : 4. If the total of the girls' marks was 175, what was Ella's result?

5. If A : B is 2 : 3, B : C is 2 : 3 and C : D is 2 : 3, what is the ratio of A : B : C : D?

6. A map has been drawn to the scale of 1 : 200 000. A national park is shown on the map with an area of 1000 km^2. What is the area of the park on the map, in square centimetres?

Answers pages 137–138

KEY SKILL

19 Ratio and rates: Rates A

HINTS

- A rate is a particular kind of ratio in which two quantities are measured in different units.
- It is often expressed as a number of the first unit in terms of one of the second unit, e.g. km/h, m/s, mL/min, L/100 km, etc.

Reminder!

- Read the question carefully to identify what needs to be found.
- Re-read the question when you've finished your answer to make sure that the question has been answered and your solution makes sense.

Examples

Josie wanted to print 24 T-shirts with a logo. She obtained quotes from three different companies. The quotes were: 6 shirts for \$24.60, 4 shirts for \$18.20, 8 shirts for \$32.40. Josie chose the best deal. How much did she pay for the T-shirts?

Solution

6-shirt deal: Cost $= 24.6 \times 4$
$= 98.4$

4-shirt deal: Cost $= 18.2 \times 6$
$= 109.2$

8-shirt deal: Cost $= 32.4 \times 3$
$= 97.2$

$\therefore$ the best deal was 8 shirts for \$32.40.

$\therefore$ the total cost was \$97.20.

FOCUS on ...

1. The question: Asks you to find the cheapest cost.
2. The information: Gives you the three quotes and the number of T-shirts to be printed.
3. Your working: For each of the deals multiply the price by a number which will give 24 shirts and the best deal is the lowest price.
   4 lots of the 6-shirt deal: cost $= 24.6 \times 4$
   $= 98.4$
   6 lots of the 4-shirt deal: cost $= 18.2 \times 6$
   $= 109.2$
   3 lots of the 8-shirt deal: cost $= 32.4 \times 3$
   $= 97.2$
   $\therefore$ best deal was 8 shirts for \$32.40 with a cost of \$97.20.
4. Your answer: Make sure you write the correct units:
   The total cost was \$97.20.

Andy travels 620 km on 50 litres of 98-octane petrol and 580 km on 50 litres of 91-octane petrol. If the price of 98-octane is 172.9 cents per litre and the 91-octane is 164.9 cents per litre, find the difference in price per 100 km.

Solution

98-octane:

Price per 100 km $= 172.9 \times 50 \div 620 \times 100$
$= 1394$ (nearest whole)

$\therefore$ price is \$13.94/100 km

91-octane:

Price per 100 km $= 164.9 \times 50 \div 580 \times 100$
$= 1422$ (nearest whole)

$\therefore$ price is \$14.22/100 km

$\therefore$ the 91-octane is 28 cents/100 km more expensive.

FOCUS on ...

1. The question: Asks you to find the difference in the cost per 100 km of using two types of petrol.
2. The information: Gives you cost per litre, and the distance a car will travel using two different types of petrol.
3. Your working: To find the price per 100 km you multiply the cost per litre by the amount of petrol, divide by the distance travelled and then multiply by 100.
   98: Price per 100 km $= 172.9 \times 50 \div 620 \times 100$
   $= 1394$ (nearest whole)
   $\therefore$ price is \$13.94/100 km
   91: Price per 100 km $= 164.9 \times 50 \div 580 \times 100$
   $= 1422$ (nearest whole)
   $\therefore$ price is \$14.22/100 km
   Subtract to find the difference: $14.22 - 13.94 = 0.28$
4. Your answer: Make sure you write the correct units:
   The 91-octane is 28 cents/100 km more expensive.

Now try these!

1. Lemons are sold in different size bags: 4 for \$2.80, 6 for \$4.10 and 10 for \$6.90. Luke needs 60 lemons. What is the smallest amount of money he can pay for the lemons?

2. Bree and Lara are friends and work at different fast-food restaurants. Bree is paid an hourly rate of \$12.50 while Lara receives \$16 per hour. Both girls need to work to earn \$500. How much longer will it take Bree than Lara to earn the money?

3. At a supermarket, bread rolls are sold in three ways. A packet of 6 costs \$3.60, a packet of 8 \$4.40 and a packet of 12 \$6.90. If Tara buys 30 bread rolls what is the smallest amount she can pay?

4. Terri travels 580 km on 55 litres of 95-octane petrol and 525 km on 55 litres of 91-octane petrol. If the price of 95-octane is 179.9 cents per litre and the 91-octane is 168.9 cents per litre, find the difference in price per 100 km.

5. A hardware store sells screws in packets of 20 for \$4.20 or packets of 100 for \$13.50. If Riley buys exactly 240 screws, what is the cheapest average cost per nail?

6. CHALLENGE Sam and Sid are saving for a cruise which costs \$3000 each. Sam has \$600 in savings and is able to save \$155 per week. Sid has \$900 and is saving at \$120 per week. Will Sam and Sid have enough money if the cruise leaves in 16 weeks? If not, by what amount will they have to increase their weekly savings? Justify your answer with calculations.

Answers page 138

KEY SKILL

20 Ratio and rates: Rates B

HINTS

- Make sure you know the **Ratio and rates Hints** from Key Skill 19 on page 52.

Reminder!

- Read the question carefully to identify what needs to be found.
- Re-read the question when you've finished your answer to make sure that the question has been answered and your solution makes sense.

Examples

A local council has set land rates at 0.64 cents per dollar of land value. How much will be the rates for a property with a value of $205 000?

Solution

Rewrite 0.64 cents to 0.0064 dollars.

$$\begin{aligned}\text{Rates} &= 0.0064 \times 205\,000 \\ &= 1312\end{aligned}$$

∴ the amount to be paid is $1312.

FOCUS on ...

1. The question: Asks you to find the cost of the land rates.
2. The information: Gives you the value of the property and the cost of the rates per dollar.
3. Your working: Change the rates from cents to dollars.
   Rewrite 0.64 cents to 0.0064 dollars.
   Multiply the rates figure by the property value.
   $$\begin{aligned}\text{Rates} &= 0.0064 \times 205\,000 \\ &= 1312\end{aligned}$$
4. Your answer: Make sure you write the correct units: The amount to be paid is $1312.

A car uses petrol at the rate of 7.4 L/100 km. How much petrol will it use to travel 440 km?

Solution

$440 \div 100 = 4.4$

$$\begin{aligned}\text{Petrol} &= 7.4 \times 4.4 \\ &= 32.56\end{aligned}$$

∴ the car uses 32.56 litres.

FOCUS on ...

1. The question: Asks you to find the amount of petrol used to travel a specified distance.
2. The information: Gives you the rate of petrol consumption per 100 km and the distance travelled.
3. Your working: Divide 440 by 100 to find the number of 100s in 440.
   $440 \div 100 = 4.4$
   Multiply the number of 100s by 7.4.
   $$\begin{aligned}\text{Petrol} &= 7.4 \times 4.4 \\ &= 32.56\end{aligned}$$
4. Your answer: Make sure you write the correct units: The car uses 32.56 litres.

Now try these!

1. This year Comeoverere Council has set land rates at 0.521 cents per dollar of land value. How much will be the rates of a property with a value of $345 000?

2. Quentin has calculated that on a recent trip his car's petrol consumption was 8.1 L/100 km. If the distance travelled was 540 km, how many litres were used?

3. Jack and Jill live in the same suburb and pay land rates to the same local council. Rates are calculated on the value of their house blocks. This year Jack's block was valued at $300 000 and he paid $1476 rates. If Jill paid $98.40 more than Jack what is the land value of her block?

4. A car uses petrol at the rate of 8.3 L/100 km. The cost of petrol is 163.9 cents/L. What will be the cost of the petrol for the car to travel 380 km? Give your answer to the nearest cent.

5. Kristie's driving holiday cost $1431.90 in petrol and she calculated that her car averaged 8.6 L/100 km. If the average price she paid for petrol during the trip was 185 cents/litre, how far did she drive?

6. CHALLENGE On a previous water bill, Ahmed was charged $167.85 for using 90 kilolitres of water. If the price per kilolitre was increased by 3.5%, how much will Ahmed pay on his next water bill if he reduces his usage by 5%?

Answers pages 138–139

KEY SKILL

21 Ratio and rates: Distance, speed and time A

HINTS

- Speed is a rate in which distance travelled is compared to the time taken,
  e.g. Travelling 80 km in 1 hour is expressed as 80 km/h.
- There are three formulae:

$$\text{Distance} = \text{Speed} \times \text{Time}$$
$$\text{Speed} = \text{Distance} \div \text{Time}$$
$$\text{Time} = \text{Distance} \div \text{Speed}$$

  e.g. Find the distance travelled by a cyclist who averages 45 km/h for 5 hours.

$$\text{Distance} = 45 \times 5 = 225$$

  ∴ the cyclist travels 225 km.

Reminder!

- Read the question carefully to identify what needs to be found.
- Re-read the question when you've finished your answer to make sure that the question has been answered and your solution makes sense.

Examples

Symon left home at 11:20 am and drove to Coolangatta at an average speed of 90 km/h. If he arrived at his destination at 5:10 pm, what distance did he travel?

Solution

Elapsed time = 11:20 am to 5:10 pm
= 5 hours 50 min
= $5\frac{5}{6}$ h

Distance = $90 \times 5\frac{5}{6}$
= 525

∴ Symon travelled 525 km.

FOCUS on ...

1. **The question:** Asks you to find the distance.
2. **The information:** Gives you the start and finish time and the average speed.
3. **Your working:** Find the time of travel.
   11:20 am to 5:10 pm = 5 hours 50 min
   = $5\frac{5}{6}$ h
   Use the formula:
   Distance = Speed × Time
   = $90 \times 5\frac{5}{6}$
   = 525
4. **Your answer:** Make sure you write the correct units: Symon travelled 525 km.

Two sisters driving separate cars start out on a journey together. Lily averages 80 km/h and Sally averages 90 km/h. When Lily has travelled 360 km, how far has Sally travelled?

Solution

Lily's time = 360 ÷ 80
= 4.5

∴ Lily has travelled for 4.5 hours.

Sally's distance = 90 × 4.5
= 405

∴ Sally has travelled 405 km.

FOCUS on ...

1. **The question:** Asks you to find the distance that Sally has travelled.
2. **The information:** Gives you the speed of both sisters and the distance Lily has travelled.
3. **Your working:** Use the formula:
   Time = Distance ÷ Speed
   = 360 ÷ 80
   = 4.5
   ∴ Lily and Sally travelled for 4.5 hours.
   Use the formula:
   Distance = Speed × Time
   = 90 × 4.5
   = 405
4. **Your answer:** Make sure you write the correct units: Sally has travelled 405 km.

Now try these!

1. A motorist left Town A at 11:45 am and drove at an average speed of 78 km/h. If he arrived at his destination at 3:15 pm what was the distance travelled?

2. A man completes a journey in 10 hours. He travels the first half of the journey at a rate of 20 km/h and the second half at a rate of 25 km/h. Find the total distance of the journey.

3. An aeroplane travels 100 km in 10 minutes. If it maintains this speed for 3 hours how far will it travel?

4. Two cars set off from Quinns Creek at 8:30 am. The first car travels at an average speed of 85 km/h and the second car travels at an average speed which is 5 km/h slower than the first car. By midday, what is the difference in the distances the cars have travelled?

5. Kaitlyn drove at an average speed of 70 km/h between 5:20 am and 8:20 am. She then travelled at an average speed of 85 km/h until 9:50 am. What was the total distance Kaitlyn travelled?

6. CHALLENGE Elise takes 10 minutes to cycle around a rectangular park at an average speed of 12 km/h. If the ratio of length to width of the park is 3 : 2, what is the area of the park in hectares?

Answers page 139

KEY SKILL

22 Ratio and rates: Distance, speed and time B

HINTS

- Speed is a rate by which distance travelled is compared to time taken, e.g. Travelling 80 km in 1 hour means 80 km/h.
- There are three formulae:

  Distance = Speed × Time

  Speed = Distance ÷ Time

  Time = Distance ÷ Speed

  e.g. Find the speed if a car travels 300 km in 6 hours.

  Speed = 300 ÷ 6

  = 50

  ∴ the average speed is 50 km/h.

Reminder!

- Read the question carefully to identify what needs to be found.
- Re-read the question when you've finished your answer to make sure that the question has been answered and your solution makes sense.

Examples

Bernie drove for 4 hours. At the beginning his car's odometer read 38 157, while at the end of the journey it read 38 409. What was his average speed?

Solution

Distance = 38 409 − 38 157

= 252

Speed = Distance ÷ Time

= 252 ÷ 4

= 63

∴ Bernie averages 63 km/h.

FOCUS on ...

1. The question: Asks you to find the average speed.
2. The information: Gives you the start and finish odometer readings and the travelling time.
3. Your working: Subtract to find the distance.

   Distance = 38 409 − 38 157

   = 252

   ∴ Bernie travelled a distance of 252 km.

   Use the formula, Speed = Distance ÷ Time:

   Speed = 252 ÷ 4

   = 63
4. Your answer: Make sure you write the correct units: Bernie averages 63 km/h.

Ravi works in a coal mine and lives in a nearby town. One evening he leaves work at 5:15 pm and arrives home at 6:00 pm. If he lives 60 km from the mine, what is his average speed for the trip?

Solution

Time = 45 minutes

= 0.75 h

Speed = Distance ÷ Time

= 60 ÷ 0.75

= 80

∴ Ravi averaged 80 km/h.

FOCUS on ...

1. The question: Asks you to find the average speed.
2. The information: Gives you the start and finish times and the distance.
3. Your working: Subtract to find the time.

   Time = 6:00 pm − 5:15 pm

   = 45 minutes

   = 0.75 h

   Use the formula, Speed = Distance ÷ Time.

   Speed = 60 ÷ 0.75

   = 80
4. Your answer: Make sure you write the correct units: Ravi averaged 80 km/h.

Now try these!

1. The odometer reading on Sara's car as she left for her eight-hour road-trip showed 63 871. When she had completed the journey the reading was 64 559. What was her average speed?

2. A bus leaves Kirkton at 4:45 pm and travels 90 km to Wyalam. If the bus arrives at Wyalam at 6:15 pm, what was the average speed of the trip?

3. The ratio between the speeds of two trains is 5 : 6. If the second train travels 360 km in 4 hours, find the average speed of the first train.

4. Every morning Alice leaves home at 7:40 am and travels 48 km to work, arriving at 9 am. This morning the traffic was very heavy and she arrived 10 minutes later than usual. Find the difference between her average speeds.

5. A plane travelled a distance of 2400 km in 3 hours with the wind. The return trip takes an hour longer. Find the speed of the plane in still air and find the speed of the wind.

6. CHALLENGE A truckdriver averages 75 km/h for the first half of a trip and then increases his average speed by 15 km/h for the second half of the trip. What is his average speed for the trip, if the total journey covers 300 km? Give your answer to the nearest kilometre per hour.

Answers pages 139–140

KEY SKILL

23 Ratio and rates: Distance, speed and time C

HINTS

- Speed is a rate by which distance travelled is compared to the time taken,
  e.g. Travelling 80 km in 1 hour means 80 km/h.
- There are three formulae:
  Distance = Speed × Time
  Speed = Distance ÷ Time
  Time = Distance ÷ Speed
  e.g. Find the time taken for a car to travel 320 km at an average speed of 80 km/h.
  Time = 320 ÷ 80
  = 4 h
  ∴ it will take 4 hours.

Reminder!

- Read the question carefully to identify what needs to be found.
- Re-read the question when you've finished your answer to make sure that the question has been answered and your solution makes sense.

Examples

Rachel leaves at 8:30 am and travels in her car at an average speed of 68 km/h. If she travels 306 km, what time does she arrive?

Solution

Time = Distance ÷ Speed
= 306 ÷ 68
= 4.5

∴ she travels for 4 h 30 min.

Now, 8:30 am plus 4 h 30 min is 1 pm.

∴ Rachel arrives at 1 pm.

FOCUS on ...

1. The question: Asks you to find the time of arrival.
2. The information: Gives you the start time, the speed and the distance.
3. Your working: Use the formula:
   Time = Distance ÷ Speed
   = 306 ÷ 68
   = 4.5
   ∴ she travelled for 4 h 30 min.
   Add this elapsed time to 8:30 am.
   8:30 am plus 4 h 30 min is 1 pm.
4. Your answer: Make sure you write the correct units:
   Rachel arrives at 1 pm.

Kynie leaves Bendigo at quarter to ten in the morning to travel 400 km to Mildura. Her average speed for the trip is 75 km/h. She needs to be in Mildura at 3 pm for an appointment. Determine whether Kynie will be early or late for her appointment.

Solution

Time = Distance ÷ Speed
= 400 ÷ 75
= $5\frac{1}{3}$

∴ she travelled for 5 h 20 min.

Now, 9:45 am plus 5 h 20 min is 3:05 pm.

∴ Kynie will arrive 5 minutes late for her appointment.

FOCUS on ...

1. The question: Asks you to find the time of arrival.
2. The information: Gives you the start time, the speed and the distance.
3. Your working: Use the formula:
   Time = Distance ÷ Speed
   = 400 ÷ 75
   = $5\frac{1}{3}$
   ∴ she travelled for 5 h 20 min.
   Add this elapsed time to 9:45 am.
   9:45 am plus 5 h 20 min is 3:05 pm.
4. Your answer: Make sure you write the correct units:
   Kynie will arrive at 3:05 pm, and so will be late for her appointment.

Now try these!

1. Liam left work at 5:20 pm and travelled home at an average speed of 64 km/h. If the distance from work to home is 32 km, what time did Liam arrive home?

2. Grace left Port Macquarie at twenty to one in the afternoon to travel 851 km to Bourke. Her average speed for the trip was 92 km/h. What time did she arrive in Bourke?

3. A cyclist rode a distance of 88 km and arrived at her destination at 1:30 pm. If she rode at an average speed of 32 km/h, what time did she start?

4. The Indian Pacific train travels along a 478-km straight section of rail across the Nullabor Plain. If the train averages 112 km/h along the section, how long will the train take between bends in the track? Give your answer to the nearest minute.

5. Larry and Pete are brothers and live in the same house. The distance from their home to Abelsford is 48 km. Larry leaves home at 6 am and walks at an average speed of 6 km/h towards Abelsford. Four and three-quarter hours later Pete leaves home and cycles at an average speed of 16 km/h towards Abelsford. Who arrives at Abelsford first?

6. CHALLENGE A train is 1 km long. It travels through a 2-km tunnel at a speed of 60 km/h. From the moment the train enters the tunnel how long will it take before the entire train travels completely through the tunnel?

Answers page 140

KEY SKILL

24 Ratio and rates: Distance, speed and time D

HINTS

- Make sure you know the **Distance, speed and time Hints** which appeared on pages 56, 58 and 60.
- If two objects are travelling in the same direction, the relative speed is found by subtracting the two speeds,
  e.g. Two cars are travelling in the same direction on the same road at 80 km/h and 65 km/h. As $80 - 65 = 15$, the relative speed of the faster car is 15 km/h.

Reminder!

- Read the question carefully to identify what needs to be found.
- Re-read the question when you've finished your answer to make sure that the question has been answered and your solution makes sense.

Examples

Kurt and Fiona leave from the same location at the same time and jog in the same direction. If Kurt jogs at an average speed of 10 km/h and Fiona at 7 km/h, how far behind Kurt is Fiona after 40 minutes?

Solution

Speed difference $= 10 - 7 = 3$

$\therefore$ difference in speed is 3 km/h.

As 40 minutes $= \frac{2}{3}$ h,

$$\text{Distance} = 3 \times \frac{2}{3}$$
$$= 2$$

$\therefore$ Fiona is 2 km behind.

FOCUS on ...

1. **The question:** Asks you to find the distance Fiona is behind Kurt after 40 minutes.
2. **The information:** Gives you the two speeds and a time of 40 minutes.
3. **Your working:** Find the difference in speeds.
   Difference $= 10 - 7$
   $= 3$
   Difference in speed is 3 km/h.
   Convert 40 minutes to hours: $\frac{40}{60} = \frac{2}{3}$ h.
   Use the formula:
   Distance = Speed × Time
   $= 3 \times \frac{2}{3}$
   $= 2$
4. **Your answer:** Make sure you write the correct units: Fiona is 2 km behind.

At 2:40 pm, two cyclists leave the same location and ride on the same bikeway in the same direction. One travels at 20 km/h and the other at 12 km/h. At what time are they 4 km apart?

Solution

Difference $= 20 - 12$
$= 8$

Difference in speed = 8 km/h

Time = Distance ÷ speed
$= 4 \div 8$
$= 0.5$

$\therefore$ after half an hour

As 2:40 pm plus 30 min is 3:10 pm, the cyclists are 4 km apart at 3:10 pm.

FOCUS on ...

1. **The question:** Asks you to find the time when the cyclists are 4 km apart.
2. **The information:** Gives you the two speeds and the distance between the cyclists.
3. **Your working:** Find the difference in speeds.
   Difference $= 20 - 12$
   $= 8$
   Difference in speed is 8 km/h.
   Use the formula:
   Distance = Distance ÷ Speed
   $= 4 \div 8$
   $= 0.5$
   Add half an hour to 2:40 pm.
   2:40 pm plus 30 min = 3:10 pm
4. **Your answer:** Make sure you write the correct units: At 3:10 pm the cyclists are 4 km apart.

Now try these!

1. Tracey leaves school and walks home at an average speed of 6 km/h. Her brother Ethan leaves school at the same time, taking the same route, but walking more quickly at 8 km/h. How far is Tracey from home when Ethan arrives home after 15 minutes? Give your answer in metres.

2. Two cars leave Springfield at 3 pm and travel on the same road to Willoughby Junction. One car travels at an average speed of 80 km/h and the other car at 92 km/h. At what time are the cars 20 km apart?

3. After work, Mr and Mrs Alder meet for dinner before driving home in separate cars at 8:55 pm. Mr Alder drives his car at an average speed of 32 km/h while his wife drives at 38 km/h. Both drivers leave at the same time and take the same route. Mrs Alder arrives at 9:35 pm. At that time, how far is her husband from home?

4. A motorist averages 80 km/h for a journey and a cyclist averages 50 km/h for the same trip. The motorist takes 1 hour 30 minutes less time to complete the trip. How far has the cyclist still to travel when the motorist finishes the journey?

5. A train is travelling at 100 km/h and catches up to a car which is travelling at 64 km/h on the road parallel to the railway track. If it takes 40 seconds for all the train to pass the driver of the car, how long is the train?

6. CHALLENGE Matthew and Blake leave Smithfield in separate cars on the same route to Antcliffe. Matthew averages 90 km/h and Blake 75 km/h. Matthew passes the sign below at 4:30 pm. At what time will Blake reach Antcliffe?

   270 Smithfield   Antcliffe 45

Answers pages 140–141

KEY SKILL

25 Ratio and rates: Distance, speed and time E

HINTS

- Make sure you know the **Distance, speed and time Hints** which appeared on pages 56, 58, 60 and 62.
- If two objects are travelling in opposite directions (either towards, or away from each other), the relative speed is found by adding the two speeds,
  e.g. Two cars are travelling away from each other on the same road at 90 km/h and 85 km/h. As 90 + 85 = 175, the relative speed of the cars is 175 km/h.

Reminder!

- Read the question carefully to identify what needs to be found.
- Re-read the question when you've finished your answer to make sure that the question has been answered and your solution makes sense.

Examples

Nathan and Brook break down on the side of a straight section of road. They decide to walk in opposite directions for help. If Nathan walks at 6 km/h and Brook at 9 km/h, how far apart are they after 50 minutes?

Solution

Relative speed $= 6 + 9$
$= 15$

As 50 min $= \frac{50}{60} = \frac{5}{6}$h,

Distance $= 15 \times \frac{5}{6}$
$= 12.5$

$\therefore$ they are 12.5 km apart.

FOCUS on ...

1. The question: Asks you to find the distance Nathan and Brook are apart after 50 minutes.
2. The information: Gives you the two speeds and a time of 50 minutes.
3. Your working: Find the relative speed.
   Relative speed $= 6 + 9$
   $= 15$
   Change the time into hours.
   50 min $= \frac{50}{60} = \frac{5}{6}$h
   Use the formula: Distance = Speed × Time
   $= 15 \times \frac{5}{6}$
   $= 12.5$
4. Your answer: Make sure you write the correct units:
   They are 12.5 km apart.

Two swimmers start at opposite ends of a 50-metre pool. Greg can swim at a speed of 1.5 m/s and Sharon can swim at 1 m/s. How far will Greg swim before he meets Sharon?

Solution

As $1.5 + 1 = 2.5$, then use a relative speed of 2.5 m/s.

Time $= 50 \div 2.5 = 20$

$\therefore$ they meet after 20 seconds.

Greg's distance $= 1.5 \times 20$
$= 30$

$\therefore$ Greg swims 30 m.

FOCUS on ...

1. The question: Asks you to find the distance Greg swims before he meets Sharon.
2. The information: Gives you the two speeds and the total distance they will swim until they meet.
3. Your working: Find the relative speed.
   Relative speed $= 1.5 + 1 = 2.5$
   Use the formula: Time = Distance ÷ Speed
   $= 50 \div 2.5$
   $= 20$
   They meet after 20 seconds.
   Find the distance Greg has swum in the 20 seconds.
   Use the formula: Distance = Speed × Time
   $= 1.5 \times 20$
   $= 30$
4. Your answer: Make sure you write the correct units:
   Greg swims 30 m.

Now try these!

1. On a calm morning at a busy airport, planes are taking off on parallel runways in opposite directions. Two planes take off simultaneously at 7:10 am. One travels at 720 km/h and the other in the opposite direction at 780 km/h. How far apart are the planes at 9:20 am?

2. At 8:50 am Harry and Sally start from places 16 km apart at the same time and walk towards each other. Harry walks at 7 km/h and Sally at 5 km/h. At what time will Harry meet Sally?

3. Sam and Laura are in Rome and leave their hotel at 7:40 am to walk on a straight section of road in opposite directions. Both girls walk at an average speed of 8 km/h. At what time are the girls 2 km apart?

4. A 340-km train line runs between Eastburg and Westburg. At 6:20 pm Train A leaves Eastburg and travels at an average speed of 80 km/h towards Westburg. At the same time, Train B leaves Westburg and travels at an average speed of 90 km/h towards Eastburg. When and where do the two trains meet?

5. Angela and Ben walk at 4 km/h and 6 km/h respectively. They start from the same point on a cross-country circuit but walk in opposite directions. If they meet in 18 minutes, what is the length of the circuit?

6. CHALLENGE Haley and Jarrod break down on a secluded road. At 3:30 pm they decide to walk for help. Haley walks to the north at 6 km/h and Jarrod walks south at 9 km/h. Their plan is to walk for 40 minutes and Haley is to stop and wait for Jarrod to jog back to her position. If Jarrod can jog at an average speed of 12 km/h, what time will he reach Haley?

Answers pages 141–142

REVISION TEST  Level of difficulty—Average

1 A local council charges rates on the basis of the value of the property. Nathan's property is valued at $220 000 and he pays rates of $1530. His neighbour Lauren pays $120 more than him. What is the value of Lauren's property?

2 A motorist was driving at a speed of 72 km/h in the rain. As the car drove under an overhead bridge the rain stopped for 0.9 seconds. How wide was the bridge?

3 Zoe left Glendonbrook and drove a distance of 84 km at an average speed of 70 km/h before stopping for a 20-minute break. She then travelled for 120 km at an average speed of 90 km/h. If she left Glendonbrook at 6:30 pm, what time did she reach her destination?

4 At 4:00 pm a train leaves a station and travels at an average speed of 52 km/h. One-and-a-half hours later a second train leaves the same station and travels on the same track at an average speed of 64 km/h. How far apart are the drivers of the trains at 9 pm?

5 Pearl walks from her home to school at an average speed of 6 km/h, and then jogs home at an average speed of 12 km/h. If her whole journey takes an hour, how far is it from home to school?

6 Mel cycled from Campbell Town to Swansea. She covered one-fifth of the trip in the first hour and one-third of the trip in the second hour. She took two hours to cycle the remaining 28 km. Find Mel's average speed for the whole trip.

Answers page 142

REVISION TEST  Level of difficulty—Challenging

1. Karen and Lesley are in a book club and are both reading a 700-page novel. By 12 March Karen has read 320 pages and is reading at 20 pages a day, while Lesley has read 380 pages and is reading at 16 pages a day. Who will have completed the book by the deadline of the end of the month?

2. Last year Mr Briar paid $1200 for land rates on his property valued at $340 000. This year the council increased the rate calculation by 2.5% and Mr Briar's property increased in value by 8%. Find the increase in the rates.

3. A train is 950 m long. It travels through a 2.5 km tunnel at an average speed of 75 km/h. From the moment the train enters the tunnel how long will it take before the entire train travels completely through the tunnel?

4. A cyclist averages 24 km/h for the first third of a trip, 18 km/h for the middle section and 15 km/h for the final third. If the total distance was 54 km, what was the time taken for the trip, in hours and minutes?

5. Angela and Hung walk at 6 km/h and 9 km/h respectively. They start from the same point on a circuit but walk in opposite directions. They pass each other in 30 minutes. How much longer will Angela take to complete the course?

6. A park is rectangular in shape with dimensions in the ratio of 5 : 3. Adam takes 28 minutes to jog around the perimeter of the park at an average speed of 11.4 km/h. The council is to fertilise the park at the rate of 20 kg/ha. What is the amount of fertiliser required? (to the nearest tonne)?

Answers pages 142–143

KEY SKILL

26 Equations A

HINTS

- An equation is an algebraic expression with an equals sign.
- When solving an equation, the aim is to finish with the pronumeral equal to a number.
- When solving equations remember to perform the same operation (add, subtract, multiply or divide) on both sides of the equation by the same number at the same time.
- To solve a problem using an equation, let the unknown be x.

Reminder!

- Read the question carefully to identify what needs to be found.
- Re-read the question when you've finished your answer to make sure that the question has been answered and your solution makes sense.

Examples

A Lotto win of $400 is to be divided between Fiona and Anne, so that Fiona receives $80 more than Anne. How much money does Anne receive?

Solution

Let Anne's amount $= x$

$\therefore$ Fiona's amount $= x + 80$

$$x + x + 80 = 400$$
$$2x + 80 = 400$$
$$2x = 400 - 80$$
$$2x = 320$$
$$x = 160$$

$\therefore$ Anne receives $160.

FOCUS on ...

1. The question: Asks you to find the amount Anne receives.
2. The information: Gives you the total amount of money and the difference between the amounts each person receives.
3. Your working: You need to introduce a pronumeral to set up an equation to solve.
   Let Anne's amount = x.
   Express Fiona's amount in terms of x.
   Fiona's amount = x + 80
   The total of these expressions equals 400.
   $$x + x + 80 = 400$$
   $$2x + 80 = 400$$
   $$2x = 400 - 80$$
   $$2x = 320$$
   $$x = 160$$
4. Your answer: Make sure you write the correct units:
   Anne receives $160.

In the last council elections there was a total of 12 400 votes for the two candidates. The winner received 2600 more votes than the loser. How many votes did each candidate receive?

Solution

Let the number of votes for the loser $= x$

$\therefore$ number of votes for the winner $= x + 2600$

$$x + x + 2600 = 12\,400$$
$$2x + 2600 = 12\,400$$
$$2x = 12\,400 - 2600$$
$$2x = 9800$$
$$x = 4900$$

$\therefore$ the loser received 4900 votes and the winner received 7500 votes.

FOCUS on ...

1. The question: Asks you to find the number of votes for each candidate.
2. The information: Gives you the total number of votes and the difference between votes for the two candidates.
3. Your working: You need to introduce a pronumeral to set up an equation to solve.
   Let loser's number of votes = x
   Express the winner's votes in terms of x.
   Winner's votes = x + 2600
   The total of these expressions equals 12 400:
   $$x + x + 2600 = 12\,400$$
   $$2x + 2600 = 12\,400$$
   $$2x = 12\,400 - 2600$$
   $$2x = 9800$$
   $$x = 4900$$
4. Your answer: Make sure you write the correct units:
   The loser received 4900 votes and the winner received 7500 votes.

Now try these!

1. A bangle and a watch altogether cost \$240. The watch costs \$70 more than the bangle. What is the cost of the bangle?

2. A pole is 8.6 m in length. If the pole is cut into two lengths where one length is 2.8 m longer than the other, find the length of each pole.

3. A television and a computer were sold for a total of \$1350. The television cost \$220 more than the computer. What was the cost of the television?

4. The distance from Albert to Pompeda is 326 km. Leighwood is on the road between the two towns and is 80 km closer to Albert than Pompeda. How far is Leighwood from Albert?

5. Sisters Lia and Mia decided to combine their money and buy a single present for their mother's birthday. They spent a total of \$120 on the present. If Mia contributed \$35 more than her sister Lia, how much did Lia contribute?

6. CHALLENGE The average of the ages of three brothers is 13. If the middle-aged boy is 9 years younger than the eldest and 3 years older than the youngest, how old are the boys?

Answers pages 143–144

KEY SKILL

27 Equations B

HINTS

- Make sure you know the **Equations Hints** from Key Skill 26 on page 68.

Reminder!

- Read the question carefully to identify what needs to be found.
- Re-read the question when you've finished your answer to make sure that the question has been answered and your solution makes sense.

Examples

Laura has half as many posters as Delvene and Josie has twice as many as Delvene. Altogether they have 42 posters. How many posters does each girl have?

Solution

Let $2x$ be the number of Delvene's posters.

Laura has x, Josie has $4x$.

$$2x + x + 4x = 42$$
$$7x = 42$$
$$x = 6$$

$\therefore$ Laura has 6 posters, Delvene has 12 posters and Josie has 24 posters.

FOCUS on …

1. The question: Asks you to find the number of each girl's posters.
2. The information: Gives you the relationship between the number of posters of each girl and the total number of posters.
3. Your working: You need to introduce a pronumeral to set up an equation to solve.
   Let Laura's number of posters = x.
   Express Josie's and Delvene's number of posters in terms of x.
   Delvene's number of posters = 2x; Josie's number of posters = 4x.
   The sum of the posters is 42.
   $$2x + x + 4x = 42$$
   $$7x = 42$$
   $$x = 6$$
4. Your answer: Make sure you write the correct units:
   Laura has 6 posters, Delvene has 12 posters and Josie has 24 posters.

Aaron's age is three times Josh's age. In 8 years from now Aaron will be twice as old as Josh. How old are the boys now?

Solution

Let Josh's age now $= x$

$\therefore$ Aaron's age now $= 3x$

In 8 years: Aaron is $3x + 8$ and Josh is $x + 8$

$$3x + 8 = 2(x + 8)$$
$$3x + 8 = 2x + 16$$
$$3x - 2x = 16 - 8$$
$$x = 8$$

$\therefore$ Josh is 8 years old and Aaron is 24 years old.

FOCUS on …

1. The question: Asks you to find the ages of the boys.
2. The information: Gives you the relationship between the boys' ages now and the relationship in 8 years' time.
3. Your working: You need to introduce a pronumeral to set up an equation to solve.
   Let Josh's age now = x.
   Let Aaron's age = 3x.
   Express the ages in 8 years' time.
   Josh is x + 8 and Aaron is 3x + 8.
   Aaron's age will equal twice Josh's age.
   $$3x + 8 = 2(x + 8)$$
   $$3x + 8 = 2x + 16$$
   $$3x - 2x = 16 - 8$$
   $$x = 8$$
4. Your answer: Make sure you write the correct units:
   Josh is 8 years old and Aaron is 24 years old.

Now try these!

1. A class survey was held to find the number of dogs, cats and rabbits kept as household pets. There were twice as many dogs as cats, and half as many rabbits as cats. If there was a total of 28 pets, how many were dogs?

2. Ken is twice as old as Kim. Ten years ago, Ken was three times Kim's age. What is the age of Ken now?

3. Serena is three times as old as Shanais. Ethan is 5 years younger than Shanais. In 6 years' time, the sum of all their ages will be 48. How old is Ethan now?

4. Hung spent $3.40 on lunch at the canteen. He bought a chicken wrap, an apple and a juice from the canteen. The cost of the juice was 20 cents more than the apple, and 60 cents less than the wrap. What was the price of the wrap?

5. A total of 120 people attended a family reunion. There were twice as many women as men and a third as many children as men. How many women were at the reunion?

6. CHALLENGE Kate's grandmother is four times as old as Kate is now. In 6 years' time she will be 2 years older than three times Kate's age. How old are Kate and her grandmother now?

Answers page 144

KEY SKILL

28 Equations and consecutive numbers

HINTS

- Make sure you know the **Equations Hints** from Key Skill 26 on page 68.
- The numbers 3, 4, 5 are consecutive.
- Consecutive even numbers and consecutive odd numbers both have a difference of 2, e.g. 6, 8, 10 or 3, 5, 7.
- If the first number is x, the next consecutive number is $x + 1$.
- If x is odd, the next consecutive odd number is $x + 2$.

Reminder!

- Read the question carefully to identify what needs to be found.
- Re-read the question when you've finished your answer to make sure that the question has been answered and your solution makes sense.

Examples

The sum of three consecutive numbers is 30. Solve an equation to find the numbers.

Solution

Let the numbers be x, $x + 1$ and $x + 2$

$$x + x + 1 + x + 2 = 30$$
$$3x + 3 = 30$$
$$3x = 30 - 3$$
$$3x = 27$$
$$x = 9$$

∴ the numbers are 9, 10 and 11.

FOCUS on ...

1. The question: Asks you to find three consecutive numbers.
2. The information: Gives you the sum of the numbers.
3. Your working: You need to introduce a pronumeral to set up an equation to solve.
   Let the smallest number = x.
   Express the other numbers in terms of x.
   Add 1 and 2 to x: numbers are x + 1, x + 2.
   The sum of the numbers is 30.
   $$x + x + 1 + x + 2 = 30$$
   $$3x + 3 = 30$$
   $$3x = 30 - 3$$
   $$3x = 27$$
   $$x = 9$$
4. Your answer: The numbers are 9, 10 and 11.

The sum of two consecutive odd numbers is 32. What is the larger number?

Solution

Let the numbers be x and $x + 2$.

$$x + x + 2 = 32$$
$$2x + 2 = 32$$
$$2x = 32 - 2$$
$$2x = 30$$
$$x = 15$$

The numbers are 15 and 17.

∴ the larger number is 17.

FOCUS on ...

1. The question: Asks you to find two consecutive odd numbers.
2. The information: Gives you the sum of the numbers.
3. Your working: You need to introduce a pronumeral to set up an equation to solve.
   Let the smaller number = x.
   Express the other number in terms of x.
   Add 2 to x: the larger number is x + 2.
   The sum of the numbers is 32.
   $$x + x + 2 = 32$$
   $$2x + 2 = 32$$
   $$2x = 32 - 2$$
   $$2x = 30$$
   $$x = 15$$
4. Your answer: The numbers are 15 and 17. The larger number is 17.

Now try these!

1. The sum of three consecutive numbers is 66. What are the numbers?

2. The sum of three consecutive odd numbers is 45. What are the numbers?

3. When added, three consecutive numbers have a sum of −24. What are the numbers?

4. Find three consecutive numbers such that the sum of the first and third numbers is 32.

5. Find three consecutive numbers such that four times the sum of the first two numbers is the same as five times the third number.

6. CHALLENGE Find three consecutive odd numbers such that five times the middle number is three more than twice the sum of the first number and the third number.

Answers pages 144–145

KEY SKILL

29 Equations and measurement

HINTS

- Make sure you know the **Equations Hints** from Key Skill 26 on page 68.
- If one number is x, then the number that is three times larger is $3x$.
- If one number is x, then the number that is 8 larger is $x + 8$.
- If one number is x, then the number that is 5 smaller is $x - 5$.

Reminder!

- Read the question carefully to identify what needs to be found.
- Re-read the question when you've finished your answer to make sure that the question has been answered and your solution makes sense.

Examples

A rectangular block of land is four times as long as it is wide. If the perimeter of the block is 70 cm, what are the dimensions of the block?

Solution

Let the width of the block be x.

Let the length of the block be $4x$.

As the perimeter is 70 cm, then the sum of the length and width is 35 cm.

$$4x + x = 35$$
$$5x = 35$$
$$x = 7$$

$\therefore$ the length is 28 cm and the width is 7 cm.

FOCUS on ...

1. The question: Asks you to find the dimensions of the rectangular block.
2. The information: Gives you the perimeter and the relationship between the length and the width.
3. Your working: You need to introduce a pronumeral to set up an equation to solve.
   Let width of block = x.
   Multiply the width by 4.
   Let the length of the block = 4x.
   The total of length and width is half the perimeter.
   $$4x + x = 35$$
   $$5x = 35$$
   $$x = 7$$
4. Your answer: Make sure you write the correct units:
   The length is 28 cm and the width is 7 cm.

The difference between the length and the breadth of a rectangular park is 31 m. If the perimeter of the park is 446 m, what is its area?

Solution

Let the width of the park be x.

Let the length of the park be $x + 31$.

As the perimeter is 446 m, then the sum of the length and breadth is 223 m.

$$2x + 31 = 223$$
$$2x = 223 - 31$$
$$2x = 192$$
$$x = 96$$

$\therefore$ dimensions are 96 m and 127 m

$$\text{Area} = 127 \times 96$$
$$= 12\,192$$

$\therefore$ area of park is 12 192 m^2.

FOCUS on ...

1. The question: Asks you to find the area.
2. The information: Gives you the difference between the length and width and the perimeter.
3. Your working: You need to introduce a pronumeral to set up an equation to solve.
   Let width of block = x.
   You add 31 because the length is 31 m longer.
   Let the length of the block = x + 31.
   The total of length and width is half the perimeter.
   $$x + 31 + x = 223$$
   $$2x + 31 = 223$$
   $$2x = 223 - 31$$
   $$2x = 192$$
   $$x = 96$$
   You add 31 to 96 to get the length.
   As 96 + 31 = 127, the length is 127 m and width is 96 m.
   Multiply to find the area.
   127 × 96 = 12 192
4. Your answer: Make sure you write the correct units:
   The area is 12 192 m^2.

Now try these!

1. The length of a rectangle is three times the width. If the perimeter is 72 cm, what are the dimensions of the rectangle?

2. The difference between the length and the width of a rectangular paddock is 16 m. If the perimeter is 160 m, find the area.

3. The length of a rectangle is 2 m longer than three times the width. If the perimeter is 100 m, what are the dimensions of the rectangle?

4. The length of a rectangular plot is 50 m more than its breadth. The cost of fencing the plot is $3.10 per metre. If the total cost of fencing is $1116, find the dimensions of the plot.

5. A paddock is in the shape of a triangle. The longest side is twice the length of the shortest side and the other side is 6 m longer than the shortest side. The cost of fencing the paddock is $2.35 per metre and the total cost is $155.10. Find the length of each side.

6. CHALLENGE A square and an equilateral triangle have the same perimeter. The sides of the square are 8 cm shorter than the sides of the triangle. What is the area of the square?

Answers page 145

KEY SKILL

30 Equations and speed

HINTS

- Speed is a rate by which distance travelled is compared to the time taken,
  e.g. Travelling 80 km in 1 hour means 80 km/h.
- There are three formulae:
  Distance = Speed × Time
  Speed = Distance ÷ Time
  Time = Distance ÷ Speed
- Make sure you know the **Equations Hints** from Key Skill 26 on page 68.

Reminder!

- Read the question carefully to identify what needs to be found.
- Re-read the question when you've finished your answer to make sure that the question has been answered and your solution makes sense.

Examples

Tom left home at 9 am and travelled to the holiday unit at an average speed of 60 km/h. His wife Megan left home 2 hours later and averaged a speed of 80 km/h. At what time did Megan catch up to Tom?

Solution

Let Tom's time of travel be x.

$\therefore$ Megan's time of travelling is $x - 2$.

As Distance = Speed × Time:

$$\begin{aligned} 60 \times x &= 80(x - 2) \\ 60x &= 80x - 160 \\ 20x &= 160 \\ x &= 8 \end{aligned}$$

9:00 am plus 8 h = 5 pm

$\therefore$ Megan caught up at 5 pm.

FOCUS on ...

1. The question: Asks you to find the time when Megan caught up to Tom.
2. The information: Gives you the speed of both Tom and Megan, the time Tom left and how much later than Tom Megan started her trip.
3. Your working: You need to introduce a pronumeral to set up an equation to solve. Let the time of Tom's trip = x. As Megan left 2 hours after Tom, her time = (x − 2). Use the formula Distance = Speed × Time. The distance travelled by both drivers will be equal.
   $$\begin{aligned} \text{Distance} = 60 \times x &= 80(x - 2) \\ 60x &= 80x - 160 \\ 20x &= 160 \\ x &= 8 \end{aligned}$$
   9:00 am plus 8 h = 5 pm
4. Your answer: Make sure you write the correct units: Megan caught up at 5 pm.

A plane leaves Tullamarine airport at 4:20 pm and flies north at an average speed of 900 km/h. A second plane leaves the same airport at 4:50 pm and flies the same route. If the second plane averages 1000 km/h, how far will it fly before it overtakes the first plane?

Solution

Let the first plane's time of flight be x.

$\therefore$ second plane's time of flight is $x - 0.5$.

As Distance = Speed × Time:

$$\begin{aligned} 900 \times x &= 1000(x - 0.5) \\ 900x &= 1000x - 500 \\ 100x &= 500 \\ x &= 5 \end{aligned}$$

$$\begin{aligned} \text{Distance of first plane} &= 900 \times 5 \\ &= 4500 \end{aligned}$$

$\therefore$ the planes will fly 4500 km.

FOCUS on ...

1. The question: Asks you to find the distance the planes have flown by the time the second plane catches up to the first.
2. The information: Gives you the speed of both planes and the times the two planes took off.
3. Your working: You need to introduce a pronumeral to set up an equation to solve. Let the time of the first plane's flight = x. As the second plane left half an hour later, its time = (x − 0.5). Use the formula Distance = Speed × Time. The distance travelled by both planes will be equal.
   $$\begin{aligned} \text{Distance} = 900 \times x &= 1000(x - 0.5) \\ 900x &= 1000x - 500 \\ 100x &= 500 \\ x &= 5 \end{aligned}$$
   Use the formula again:
   $$\begin{aligned} \text{Distance} &= 900 \times 5 \\ &= 4500 \end{aligned}$$
4. Your answer: Make sure you write the correct units: The planes will fly 4500 km.

Now try these!

1 Layla left work at 5:00 pm and drove towards her home at an average speed of 45 km/h. Ten minutes later Ava left work and followed Layla's route, averaging 54 km/h. At what time did Ava catch up to Layla?

2 Two cyclists rode along the same cycleway. Caleb started at 6:30 am and averaged 18 km/h. Jax waited 20 minutes and then cycled after Caleb at an average speed of 24 km/h. How far did Caleb ride before he was overtaken by Jax?

3 A car passes an intersection travelling west at 85 km/h. A second car passes the same intersection 45 minutes later heading west travelling at 100 km/h. How long will it take the second car to overtake the first?

4 James and Ben are in a race over a distance of 5000 m. James starts and averages 10 km/h. Five minutes later Ben starts and averages 12 km/h. Who wins the race?

5 Two friends had a race. Brian started the race and ran at an average speed of 8 km/h. Lucas waited 6 minutes and then ran at an average speed of 12 km/h. How many minutes after Lucas started did he pass Brian?

6 CHALLENGE Jocelyn left the post office at Kalgoorlie at 10 am and travelled east at an average speed of 90 km/h. An hour later Chloe left the post office and travelled west at an average speed of 100 km/h. At what time were they 356 km apart?

Answers page 146

KEY SKILL

31 Equations and mixtures

HINTS

- Make sure you know the **Equations Hints** from Key Skill 26 on page 68.
- The sum of the numbers a and $(b - a)$ is b, e.g. Two numbers add to 20. If one number is x, the other number is $20 - x$.

Reminder!

- Read the question carefully to identify what needs to be found.
- Re-read the question when you've finished your answer to make sure that the question has been answered and your solution makes sense.

Examples

A school held an out-of-uniform day to raise funds for leukemia research. 820 students made a gold coin donation of either a \$1 or \$2 coin. If \$1500 was raised, how many students donated a \$2 coin?

Solution

Let the number of \$1 coins be x.

$\therefore$ number of \$2 coins was $820 - x$.

$$\begin{aligned} 1 \times x + 2(820 - x) &= 1500 \\ x + 1640 - 2x &= 1500 \\ 1640 - x &= 1500 \\ x &= 1640 - 1500 \\ x &= 140 \end{aligned}$$

$820 - 140 = 680$

$\therefore$ 680 students each gave a \$2 coin.

FOCUS on ...

1. The question: Asks you to find the number of students who paid a \$2 coin.
2. The information: Gives you the number of students who paid a gold coin and the total amount paid.
3. Your working: You need to introduce a pronumeral to set up an equation to solve. Let the number of \$1 coins = x. This means the quantity of \$2 coins = 820 − x. Adding the values of the \$1 and \$2 coins gives 1500.
$$\begin{aligned} 1 \times x + 2(820 - x) &= 1500 \\ x + 1640 - 2x &= 1500 \\ 1640 - x &= 1500 \\ x &= 1640 - 1500 \\ x &= 140 \end{aligned}$$
As, 820 − 140 = 680, then there were 140 \$1 and 680 \$2.
4. Your answer: Make sure you write the correct units: 680 students each donated a \$2 coin.

A shopkeeper sells bags of mixed nuts. She wants to mix cashews and almonds where cashews cost \$4/kg and almonds \$8/kg. She plans to sell 100 kg of the mix. How many kilograms of each type of nut should be mixed if each mixed bag of nuts costs the shopkeeper \$5/kg?

Solution

Let the mass of cashews be x.

$\therefore$ mass of almonds is $100 - x$.

$$\begin{aligned} 4 \times x + 8(100 - x) &= 5 \times 100 \\ 4x + 800 - 8x &= 500 \\ 4x &= 300 \\ x &= 75 \end{aligned}$$

$\therefore$ 75 kg of cashews and 25 kg of almonds should be mixed.

FOCUS on ...

1. The question: Asks you to find the mass of each type of nut.
2. The information: Gives you the cost of each nut, the total mass of the nuts and the cost of each bag of nuts.
3. Your working: You need to introduce a pronumeral to set up an equation to solve. Let the mass (in kg) of cashews = x.
This means the mass (in kg) of almonds = 100 − x. Multiply the cost of each nut by its mass and add. This will equal the total cost of all nuts.
$$\begin{aligned} 4 \times x + 8(100 - x) &= 5 \times 100 \\ 4x + 800 - 8x &= 500 \\ 800 - 4x &= 500 \\ 4x &= 300 \\ x &= 75 \end{aligned}$$
Also, 100 − 75 = 25.
4. Your answer: Make sure you write the correct units: 75 kg of cashews and 25 kg of almonds should be mixed.

Now try these!

1. Tickets for a theatre group's play cost $12 for adults and $8 for children. At one performance, 128 tickets were sold and a total of $1368 was collected. How many adult tickets were sold?

2. A 50-kg wild bird seed mix is made by combining millet seed costing $1.80/kg with sunflower seeds costing $2.30/kg. How many kilograms of millet seed are needed to make a mixture that costs $2.00/kg?

3. Mrs McDonald had a farm and on her farm she had chickens and goats. The animals had a total of 66 heads and 188 legs. How many chickens and goats were on her farm?

4. A metal worker needs to make 4 kg of an alloy which contains 60% silver. He melts two alloys together: one contains 30% silver and the other 80% silver. What mass of each alloy will he need to use?

5. Leo is a wood-turning stool-maker. He makes 3-legged and 4-legged stools. In one month he used 380 legs to produce 120 stools. How many 3-legged stools did he make?

6. CHALLENGE Cooper had $10 000 to invest. He invested an amount at Sumbank at 8% interest per annum and the rest at another bank at 7% per annum. If his total annual interest was $765, find the amount invested at Sumbank.

Answers pages 146–147

KEY SKILL

32 Using formulae

HINTS

- A formula is a special type of an equation that shows the relationship between different variables.
- In $A = 3p - q$, independent variables are p and q and A is the dependent variable.
- The subject of a formula is the single variable (usually on the left-hand side), that everything else is equal to. 'To change the subject' means to rearrange the formula so that another variable is the subject of the formula.
- Formulae are used by substituting numbers for pronumerals, e.g. If $v = u + at$, find v if $u = 6$, $a = 10$, $t = 3$.

$$\begin{aligned} v &= 6 + 10 \times 3 \\ &= 36 \end{aligned}$$

Reminder!

- Read the question carefully to identify what needs to be found.
- Re-read the question when you've finished your answer to make sure that the question has been answered and your solution makes sense.

Examples

The formula $d = 4.9t^2$ is used to find the distance d, in metres, an object falls in t seconds, under the effect of gravity.
How far does Lance fall if he jumps out of a plane and waits 20 seconds before he opens his parachute?

Solution

Subst. $t = 20$ in $d = 4.9t^2$

$$\begin{aligned} &= 4.9 \times (20)^2 \\ &= 1960 \end{aligned}$$

$\therefore$ Lance falls 1960 m.

FOCUS on ...

1. The question: Asks you to find the distance Lance falls.
2. The information: Gives you the formula linking distance and time and the time (t) to be used to find the distance (d).
3. Your working: Subs. the value of t into the formula.
   $$\begin{aligned} d &= 4.9t^2 \\ &= 4.9 \times (20)^2 \\ &= 1960 \end{aligned}$$
4. Your answer: Make sure you write the correct units: Lance falls 1960 m.

The volume V, of a rectangular pyramid is found using the formula $V = \frac{lbh}{3}$, where l = length, b = breadth and h = height.
Find the height of a rectangular pyramid with length 15 m, breadth 12 m and a volume of 360 m^3.

Solution

$$\begin{aligned} V &= \frac{lbh}{3} \\ 360 &= \frac{15 \times 12 \times h}{3} \\ 360 &= 60h \\ h &= \frac{360}{60} \\ &= 6 \end{aligned}$$

$\therefore$ the height is 6 m.

FOCUS on ...

1. The question: Asks you to find the height.
2. The information: Gives you the formula linking volume with length, breadth and height. The volume (V), length (l) and breadth (b) are used to find the height (h).
3. Your working: Subs. the value of V, l and b into the formula.
   $$\begin{aligned} V &= \frac{lbh}{3} \\ 360 &= \frac{15 \times 12 \times h}{3} \\ 360 &= 60h \\ h &= \frac{360}{60} \\ &= 6 \end{aligned}$$
4. Your answer: Make sure you write the correct units: The height is 6 m.

Now try these!

1. The formula $I = Prn$ is used to find the amount of simple interest when an amount of money (P) is invested at an interest rate (r%, expressed as a decimal) for a period of time (n). Find the amount of simple interest if \$12 000 is invested at 4% per annum for 8 years.

2. The formula $A = \frac{1}{2}bh$ is used to find the area (A) of a triangle with base length b and height h. Find the length of the base of a triangle with area 120 cm^2 and height 16 cm.

3. The formula to find the cost (C) in dollars to produce rocking horses (r) is given by the formula $C = 480 + 120r$. Find the cost of producing 8 rocking horses.

4. The formula $V = \frac{4}{3}\pi r^3$ is used to find the volume (V) of a sphere with radius r. What is the radius of a sphere with a volume of 907.78 cm^3?

5. The formula $A = \frac{h}{2}(a + b)$ is used to find the area (A) of a trapezium with height h and lengths of the parallel sides a and b. What is the area of a trapezium with a height of 14 cm and parallel sides 12 cm and 18 cm?

6. CHALLENGE The formula $I = \frac{m}{h^2}$ is used to find the Body Mass Index (I) for adults with mass m kg and height h metres. Scott is 40 years old and in three months his BMI dropped by 2.26 to 28.4. If Scott is 176 cm tall, what was his previous mass?

Answers page 147

KEY SKILL

33 Linear relationships

HINTS

- A linear relationship between two variables occurs when one variable is directly proportional to another.
- When a linear relationship is graphed it appears as a straight line.
- The relationship is written as an equation and can be determined using a table or a number plane graph,

  e.g. Use the table to find the rule that links y in terms of x:

  The rule is $y = 3x + 4$.

x	0	1	2	3	4
y	4	7	10	13	16

Reminder!

- Read the question carefully to identify what needs to be found.
- Re-read the question when you've finished your answer to make sure that the question has been answered and your solution makes sense.

Examples

A tap is dripping at a constant rate and the quantity of water wasted in litres (Q) is recorded over a period of time (t) in minutes.

t	30	60	90	120	150
Q	5	10	15	20	25

Form an equation for Q in terms of t and use the formula to find the amount of water wasted after one day.

Solution

$Q = \frac{t}{6}$

24 hours = 1440 minutes

Subs. $t = 1440$ in $Q = \frac{t}{6} = \frac{1440}{6} = 240$

$\therefore$ 240 litres was wasted.

FOCUS on...

1. The question: Asks you to write the rule that links Q and t, and use this equation to find the water (Q) wasted after one day (t).
2. The information: Gives you a series of values of t and the resultant value of Q.
3. Your working: You need to use the table to find a rule linking Q and t.
   The value of Q is the value of t divided by 6:
   $Q = \frac{t}{6}$
   Now, change time to minutes: $24 \times 60 = 1440$.
   Subs. $t = 1440$ in $Q = \frac{t}{6} = \frac{1440}{6} = 240$
4. Your answer: Make sure you write the correct units:
   240 litres of water was wasted.

A length of pipe is sold according to its length and the following table details the cost ($\$C$) of different lengths ($d$, in metres).

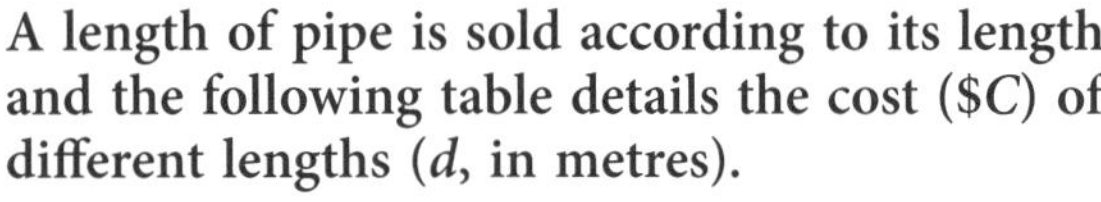

d	3	5	8	20	100
C	4.8	8	12.8	32	160

Form an equation for C in terms of d and use the formula to find the length of pipe that can be purchased for $60.80.

Solution

$C = 1.6d$

Subs. $C = 60.8$ in $C = 1.6d$.

$60.8 = 1.6d$

$d = \frac{60.8}{1.6} = 38$

$\therefore$ a 38-metre length of pipe can be purchased.

FOCUS on...

1. The question: Asks you to write the rule that links C and d, and use this equation to find the distance (d) for a cost (C) of $60.80.
2. The information: Gives you a series of values of d and the resultant value of C.
3. Your working: You need to use the table to find a rule linking C and d.
   The value of C is the value of d multiplied by 1.6.
   $C = 1.6d$
   Subs. $C = 60.8$ in $C = 1.6d$.
   $60.8 = 1.6d$
   $d = \frac{60.8}{1.6} = 38$
4. Your answer: Make sure you write the correct units:
   A 38-metre length of pipe can be purchased.

Now try these!

1. The cost (C), in dollars, of different lengths of material (n) in metres, is listed in the table.

n	2	3	5	8	10	15
C	24	36	60	96	120	180

Form an equation for C in terms of n and use the formula to find the cost of 20 m of material.

2. Megan uses the following table to advertise the charge (C), in dollars, of different periods of her babysitting (t), in hours.

t	1	2	3	5
C	25	35	45	65

Form an equation for C in terms of t and use the formula to find the cost of Megan's babysitting for 4 hours.

3. Charlotte buys 50 scented soaps to place in gift baskets she is assembling for her market stall. The table shows the relationship between the number of gift baskets (n) and the number of soaps (s) she has remaining.

n	0	2	5	7	10	12
s	50	44	35	29	20	14

Write an equation linking s and n, and find the number of soaps she has remaining after making 16 baskets.

4. The cost (C), in dollars, of hiring a taxi to travel different distances (d), in kilometres, is detailed in the table.

d	0	3	5	10
C	3.50	9.92	14.20	24.90

Form an equation for C in terms of d and use the formula to find the cost of hiring the taxi for a trip of 32 km.

5. Noah is travelling between two cities and the table shows the distance d, in kilometres, he still has to travel after driving for n hours. Noah is able to travel at 95 km/h for the entire journey.

n	1	2	4	6
d	705	610	420	230

Form an equation for d in terms of n and use the formula to find the distance he still has to travel after driving for 7 hours 30 minutes.

6. CHALLENGE Sophie ran 3 km on the first day and on each following day increased the distance by 500 m. Complete the table to show the relationship between days (n) and distance (D), in kilometres.

n	1	2	4	10
D				

Form an equation for D in terms of n and use the formula to find the distance Sophie ran on the 15th day.

Answers pages 147–148

REVISION TEST  Level of difficulty—Average

1. A tablet and a phone were sold for a total of $990. The tablet cost $200 less than the phone. What was the cost of the phone?

2. Today, Nigel is twice as old as Jack. Four years ago Nigel was three times as old as Jack. How old will Nigel be in another 5 years?

3. On a test of 20 questions, each correct answer earns 5 marks, an incorrect answer scores −3 and any question not attempted scores 0. Liam scored 53 and did not answer 3 questions. How many questions were answered correctly?

4. The formula $S = 2\pi rh + 2\pi r^2$ is used to find the surface area (S) of a closed cylinder with radius r and height h. Find the surface area of a cylinder with radius 8 cm and height 10 cm. Leave your answer correct to 2 decimal places.

5. Maureen left Alice Springs at 8 am and drove south at a steady 80 km/h. Leif left Alice Springs two hours later and drove south on the same road at 110 km/h. If both drivers maintained their speed, what time did Leif catch up to Maureen?

6. The length of a rectangle is 8 m longer than twice its width. If the perimeter is 136 m, find the area of the rectangle.

Answers page 148

REVISION TEST  Level of difficulty—Challenging

1. Theo and his grandfather Barry were comparing their ages. At present, the sum of their ages is 81. In 4 more years Barry will be six times as old as Theo was last year. How old is each person now?

2. Find three consecutive even numbers such that three times the middle number is 18 less than three times the sum of the first number and the third number.

3. Sasha hires an office to run a tutoring business. Her costs per week total $600 and she charges $50 per hour for tutoring. By using profit (P) in dollars and number of hours tutoring (n), complete the table below.

n	10	12	20	30
P				

Write a formula linking P and n and find the profit if she tutors for 28 hours.

4. Adam played in a school cricket tournament. In one match he scored a total of 90 runs, but only hit fours and sixes. If he scored three times as many fours as sixes, how many sixes did he hit?

5. The ratio of Libby's money to Grace's money was 7 : 3. After Libby spent $24 and Grace's mother gave Grace $40, both girls had the same amount of money. How much did each have at the beginning?

6. Carter left Winton at 2 pm and travelled north-west at an average speed of 80 km/h. Alyssa left Winton at 2:15 pm and drove south-east at an average speed of 100 km/h. After they had driven the same distance, Carter stopped and waited as Alyssa turned around and drove back towards him. If Alyssa maintained her previous speed, at what time did she reach Carter?

Answers pages 148–149

KEY SKILL

34 Circumference A

HINTS

- The circumference is the perimeter of a circle.
- The ratio of circumference to diameter equals pi (π). This means π = cicumference ÷ diameter
- If C = circumference, d = diameter, r = radius, then: $C = \pi d$; or $C = 2\pi r$.

Reminder!

- Read the question carefully to identify what needs to be found.
- Re-read the question when you've finished your answer to make sure that the question has been answered and your solution makes sense.

Examples

A bicycle wheel has a diameter of 62 cm. How far will the bike move in one revolution of the wheel? Give your answer to 2 decimal places.

Solution

Diameter = 62 cm

Circumference $= \pi d$

$= \pi \times 62$

$= 194.778\,7445 \ldots$

$= 194.78$ (2 dec. pl.)

∴ the bike travels 194.78 cm.

FOCUS on…

1. The question: Asks you to find the distance the wheel moves in one revolution.
2. The information: Gives you the diameter of the wheel.
3. Your working: Find the circumference.
   $C = \pi d$
   $= \pi \times 62$
   $= 194.778\,7445 \ldots$
   Correct your answer to 2 decimal places.
   $= 194.78$ (2 dec. pl.)
4. Your answer: Make sure you write the correct units:
   The bike travels 194.78 cm.

A circular running track has been marked in lanes. Two athletes P and Q run around the track once. How much further does Q run? Give your answer to the nearest metre.

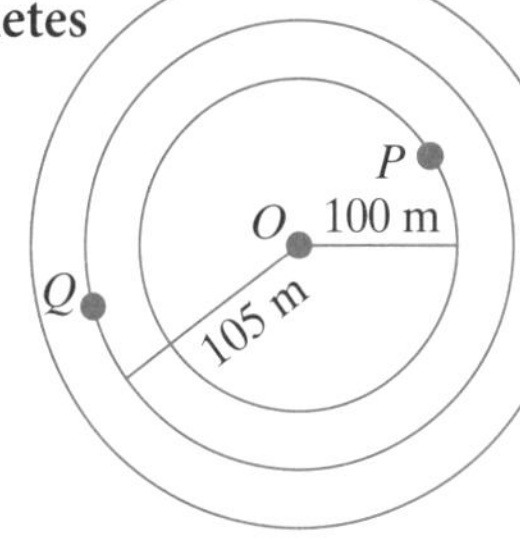

Solution

Q: $r = 105$,

$C = 2\pi r$

$= 2 \times \pi \times 105$

$= 659.734\,4573 \ldots$

P: $r = 100$,

$C = 2\pi r$

$= 2 \times \pi \times 100$

$= 628.318\,5307$

Difference $= 31.415\,926\,54 \ldots$

$= 31$ (nearest whole)

∴ Q runs 31 m further.

FOCUS on…

1. The question: Asks you to find the difference between the distances run by the two athletes.
2. The information: Gives you the radii of two circles.
3. Your working: Find the circumference of the larger circle.
   $C = 2\pi r$
   $= 2 \times \pi \times 105$
   $= 659.734\,4573\ldots$
   Find the circumference of the smaller circle.
   $C = 2 \times \pi \times 100$
   $= 628.318\,5307\ldots$
   Subtract the two circumferences.
   Difference $= 31.415\,926\,54\ldots$
   Correct your answer to the nearest whole.
   $= 31$ (nearest whole)
4. Your answer: Make sure you write the correct units:
   Q runs 31 m further than P.

Now try these!

1. Barbara has a set of 8 circular dinner plates which have gold trim around the edge of each plate. If the diameter of each plate is 30 cm, what is the total length of gold trim, to the nearest centimetre?

2. A circular track has an inside lane of radius 55 m. The outside lane has a radius of 57 m. In one circuit how much further does the runner in the outside lane have to run, to the nearest metre?

3. A circle with radius 8 cm is bent into a square. What is the length of each side of the square, to the nearest millimetre?

4. A horse is tied to a pole in the ground with a rope which is 3.2 m long. If the horse exercises by jogging in a circle around the pole, how far does the horse run in 50 revolutions? Give your answer to the nearest metre.

5. A quadrant of a circle with a radius of 8 m is used for a garden plot. How much is the cost of edging the garden if edging costs $6.30/m?

6. CHALLENGE On a clock the length of the hour hand is 3 cm shorter than the minute hand. How much further does the minute hand travel compared to the hour hand in 12 hours if the hour hand is 5 cm in length. Give your answer to the nearest centimetre.

Answers pages 149–150

KEY SKILL

35 Circumference B

HINTS

- Make sure you know the **Circumference Hints** from Key Skill 34 on page 86.

Reminder!

- Read the question carefully to identify what needs to be found.
- Re-read the question when you've finished your answer to make sure that the question has been answered and your solution makes sense.

Examples

A circular mousepad has a circumference of 68 cm. What is the diameter of the mousepad? Give your answer to 2 decimal places.

Solution

$$C = \pi d$$
$$68 = \pi \times d$$
$$d = 68 \div \pi$$
$$= 21.645\,072\,26 \ldots$$
$$= 21.65 \text{ (2 dec. pl.)}$$

$\therefore$ the diameter is 21.65 cm

FOCUS on ...

1. The question: Asks you to find the diameter of a circle if the circumference is known.
2. The information: Gives you the circumference.
3. Your working: Use the formula for the circumference.
$$C = \pi d$$
$$68 = \pi d$$
$$d = 68 \div \pi$$
$$= 21.645\,072\,26\ldots$$
Correct your answer to 2 decimal places.
$$d = 21.65 \text{ (2 dec. pl.)}$$
4. Your answer: Make sure you write the correct units: The diameter is 21.65 cm.

Lana bent a length of wire into a square with side length of 12 cm. She then reshaped the wire into a circle. What is the radius of her circle? Express the answer to 2 decimal places.

Solution

Square: $P = 4s$
$$= 4 \times 12$$
$$= 48$$

$\therefore$ the length of wire is 48 cm.

Circle: $C = 2\pi r$
$$48 = 2 \times \pi \times r$$
$$r = 48 \div 2\pi$$
$$= 48 \div (2 \times \pi)$$
$$= 7.639\,437\,268 \ldots$$
$$= 7.64 \text{ (2 dec. pl.)}$$

$\therefore$ the radius is 7.64 cm.

FOCUS on ...

1. The question: Asks you to find the radius of the formed circle.
2. The information: Gives you the side length of the square.
3. Your working: Find the perimeter of the square.
Square: $P = 4s$
$$= 4 \times 12$$
$$= 48$$
$\therefore$ the length of wire is 48 cm.
Use the circumference formula to find the radius.
Circle: $C = 2\pi r$
$$48 = 2 \times \pi \times r$$
$$r = 48 \div 2\pi$$
$$= 48 \div (2 \times \pi)$$
$$= 7.639\,437\,268 \ldots$$
Correct your answer to 2 decimal places.
$$r = 7.64 \text{ (2 dec. pl.)}$$
4. Your answer: Make sure you write the correct units: The radius is 7.64 cm.

Now try these!

1. A circular rim has a circumference of 68π cm. What is the radius of the rim, in terms of π?

2. A rectangle measuring 12 cm by 10 cm is bent into the shape of a circle. What is the diameter of the circle, to the nearest centimetre?

3. A can of beans falls from the pantry onto the kitchen floor. It then rolls a distance of 1.1 m. If it rolls exactly 5 revolutions, what is the diameter of the can, to the nearest centimetre?

4. A coin is rolled on its edge for a distance of 1 metre. If the diameter of the coin is 28 mm, how many revolutions of the coin does it roll, correct to 2 decimal places?

5. Jay is riding around a circular track at an average speed of 24 km/h for 20 minutes. How many revolutions of the track does he complete if the diameter is 80 m?

6. CHALLENGE A penny-farthing bike is built with a front wheel radius of 0.72 m and a rear wheel diameter of 0.36 m. If the bike travels one kilometre, how many more revolutions does the rear wheel do than the front wheel, to the nearest whole?

Answers page 150

KEY SKILL

36 Area of quadrilaterals and triangles A

HINTS

- The area is a measure of the space inside a plane shape.
- The formulae are:

Square: $A = s^2$ Rectangle: $A = lb$ Triangle: $A = \frac{1}{2}bh$

Parallelogram: $A = bh$ Trapezium: $A = \frac{1}{2}h(a + b)$

Rhombus/Kite: $A = \frac{1}{2}$ product of diagonals

Reminder!

- Read the question carefully to identify what needs to be found.
- Re-read the question when you've finished your answer to make sure that the question has been answered and your solution makes sense.

Examples

A rectangle and a square have the same areas. The length of a rectangle is 27 cm and each side of the square measures 18 cm. What is the breadth of the rectangle?

Solution

$$\begin{aligned}\text{Area of square} &= 18 \times 18\\ &= 324\\ \text{Breadth of rectangle} &= 324 \div 27\\ &= 12\end{aligned}$$

$\therefore$ the breadth is 12 cm.

FOCUS on ...

1. **The question:** Asks you to find the breadth of the rectangle.
2. **The information:** Gives you the length of the rectangle and the side length of a square with identical area.
3. **Your working:** Find the area of the square.
$$\begin{aligned}A &= s^2\\ &= 18^2\\ &= 324\end{aligned}$$
Substitute into the area of rectangle formula.
$$\begin{aligned}A &= lb\\ 324 &= 27 \times b\\ b &= 324 \div 27\\ &= 12\end{aligned}$$
4. **Your answer:** Make sure you write the correct units: The breadth is 12 cm.

A triangle with a base length of 32 cm has the same area as a parallelogram with base 24 cm and height 10 cm. What is the perpendicular height of the triangle?

Solution

$$\begin{aligned}\text{Area of parallelogram} &= 24 \times 10\\ &= 240\\ \text{Area of triangle} &= \frac{1}{2}bh\\ 240 &= \frac{1}{2} \times 32 \times h\\ 16h &= 240\\ h &= 240 \div 16\\ &= 15\end{aligned}$$

$\therefore$ the perpendicular height is 15 cm.

FOCUS on ...

1. **The question:** Asks you to find the perpendicular height of the triangle.
2. **The information:** Gives you the base length of the triangle and the dimensions of a parallelogram with identical area.
3. **Your working:** Find the area of the parallelogram.
$$\begin{aligned}A &= bh\\ &= 24 \times 10\\ &= 240\end{aligned}$$
Substitute into the area of triangle formula.
$$\begin{aligned}A &= \frac{1}{2}bh\\ 240 &= \frac{1}{2} \times 32 \times h\\ h &= 240 \div 16\\ &= 15\end{aligned}$$
4. **Your answer:** Make sure you write the correct units: The perpendicular height is 15 cm.

Now try these!

1. A rectangle and a square have the same areas. The width of the rectangle is 14 cm and each side of the square is 28 cm. What is the length of the rectangle?

2. A kite has diagonals of 12 cm and 16 cm. The area of the kite is the same as the area of a triangle with base 24 cm. What is the height of the triangle?

3. A parallelogram and a square have the same area. The length of the parallelogram is three times its height. If the square has a side of 6 cm, find the height of the parallelogram.

4. Rob has two paddocks on his farm that are to be fertilised at the same rate. The cost of fertilising the square paddock is $135.20. A second paddock in the shape of a rectangle costs $96. If the dimensions of the second paddock are 64 m by 30 m, what is the perimeter of the first paddock?

5. A tank is in the shape of a rectangular prism. The tank is 6 m long, 3 m wide and 2 m deep. The cost of painting a sealant on the walls and the base is $5.60 per square metre. What is the total cost of the painting?

6. CHALLENGE *ABCD* is a parallelogram and *M* lies on *DC* where *DM* = *MC*. What is the ratio of the area of triangle *DBM* to the area of the parallelogram *ABCD*?

Answers page 151

KEY SKILL

37 Area of quadrilaterals and triangles B

HINTS

- Make sure you know the **Area of quadrilaterals and triangles Hints** from Key Skill 36 on page 90.

Reminder!

- Read the question carefully to identify what needs to be found.
- Re-read the question when you've finished your answer to make sure that the question has been answered and your solution makes sense.

Examples

A parallelogram has a height of 6 cm. It has the same area as a triangle with base 18 cm and height of 16 cm. What is the length of the base of the parallelogram?

Solution

$$\text{Area of triangle} = \frac{1}{2} \times 18 \times 16 = 144$$

$$\text{Area of parallelogram} = bh$$

$$144 = b \times 6$$

$$6b = 144$$

$$b = 144 \div 6 = 24$$

$\therefore$ the length of the base is 24 cm.

FOCUS on...

1. The question: Asks you to find the base length of the parallelogram.
2. The information: Gives you the dimensions of a triangle with the same area as the parallelogram and the height of the parallelogram.
3. Your working: Find the area of the triangle.
   $$A = \frac{1}{2}bh = \frac{1}{2} \times 18 \times 16 = 144$$
   Use the formula for the area of a parallelogram.
   $$A = bh$$
   $$144 = b \times 6$$
   $$6b = 144$$
   $$b = 144 \div 6 = 24$$
4. Your answer: Make sure you write the correct units: The length of the base is 24 cm.

A rhombus has diagonals of length 12 cm and 10 cm. It has the same area as a trapezium with a height of 6 cm. What is the sum of the lengths of the two parallel sides of the trapezium?

Solution

$$\text{Area of rhombus} = \frac{1}{2} \times \text{product of diagonals}$$

$$= \frac{1}{2} \times 12 \times 10 = 60$$

$$\text{Area of trapezium} = \frac{1}{2}h(a + b)$$

$$60 = \frac{1}{2} \times 6 \times (a + b)$$

$$3 \times (a + b) = 60$$

$$a + b = 60 \div 3 = 20$$

$\therefore$ the sum of the parallel sides is 20 cm.

FOCUS on...

1. The question: Asks you to find the sum of the lengths of the parallel sides of a trapezium.
2. The information: Gives you the dimensions of a rhombus with the same area as the trapezium and the height of the trapezium.
3. Your working: Find the area of the rhombus.
   $$A = \frac{1}{2} \times \text{product of diagonals} = \frac{1}{2} \times 12 \times 10 = 60$$
   Use the formula for the area of a trapezium.
   $$A = \frac{1}{2}h(a + b)$$
   $$60 = \frac{1}{2} \times 6(a + b)$$
   $$60 = 3(a + b)$$
   $$a + b = 60 \div 3 = 20$$
4. Your answer: Make sure you write the correct units: The sum of the parallel sides is 20 cm.

Now try these!

1. The base of a parallelogram is 10 cm. The parallelogram has the same area as a rhombus with diagonals 8 cm and 20 cm. What is the perpendicular height of the parallelogram?

2. A trapezium has a perpendicular height of 10 cm. It has the same area as a rhombus with diagonals 12 cm and 15 cm. What is the sum of the two parallel sides in the trapezium?

3. A triangle has a perpendicular height of 12 cm and has the same area as a kite with diagonals 15 cm and 8 cm. What is the length of the base of the triangle?

4. A trapezium and a parallelogram have the same area and are to be covered with material which costs \$3.65 per m^2 . The total cost of covering the two shapes is \$350.40. If the height of the parallelogram is 4 m, what is its base?

5. The perimeter of a square is 24 cm. It has the same area as a kite with one diagonal 8 cm. What is the length of the other diagonal?

6. CHALLENGE Two parks of equal area are to be fertilised using fertiliser at the rate of 160 kg/ha. One park is in the shape of a parallelogram and requires 2.4 tonnes of fertiliser. The other park is in the shape of a rhombus and the length of one diagonal of the park is 1 km long. What is the length of the other diagonal?

Answers pages 151–152

KEY SKILL

38 Area of circles A

HINTS

- The area is a measure of the space inside a plane shape.
- The formula for area of a circle is: $A = \pi r^2$.

Reminder!

- Read the question carefully to identify what needs to be found.
- Re-read the question when you've finished your answer to make sure that the question has been answered and your solution makes sense.

Examples

The spray from a circular lawn sprinkler has a diameter of 6 m. What area of lawn is watered by the sprinkler? Give your answer to 2 decimal places.

Solution

Diameter = 6 m, ∴ Radius = 3 m

$A = \pi r^2$

$= \pi \times 3^2$

$= 28.274\,333\,88 \ldots$

$= 28.27$ (2 dec. pl.)

∴ the area of lawn is 28.27 m^2.

FOCUS on ...

1. **The question:** Asks you to find the area of the circle.
2. **The information:** Gives you the diameter.
3. **Your working:** Find the radius of the circle.
   As $6 \div 2 = 3$, the radius is 3 cm.
   Use the formula for the area of a circle.
   $A = \pi r^2$
   $= \pi \times 3^2$
   $= 28.274\,333\,88\ldots$
   Correct your answer to 2 decimal places.
   $= 28.27$ (2 dec. pl.)
4. **Your answer:** Make sure you write the correct units: The area is 28.27 m^2.

Chris used two sheets of paper to cut out two circles. The radius of the blue circle is 8 cm and the radius of the white circle is 12 cm. When Chris placed the blue circle on top of the white circle, what area of white circle could be seen? Leave your answer in terms of π.

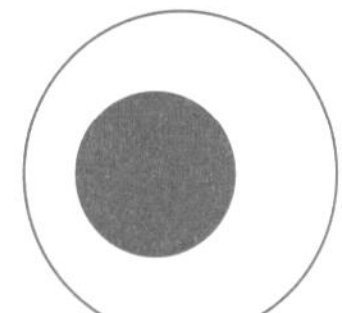

Solution

Area $= \pi \times 12^2 - \pi \times 8^2$

$= 144\pi - 64\pi$

$= 80\pi$

∴ the area of white circle shown is 80π cm^2.

FOCUS on ...

1. **The question:** Asks you to find the area of the white circle that can be seen.
2. **The information:** Gives you the radii of both circles.
3. **Your working:** Subtract the area of the small circle from the bigger.
   Area $= \pi \times 12^2 - \pi \times 8^2$
   $= 144\pi - 64\pi$
   $= 80\pi$
4. **Your answer:** Make sure you write the correct units: The area of white circle shown is 80π cm^2.

Now try these!

1. A goat is attached to a rope and the rope is tied to a post. If the rope is 3.2 m long what area of grass can be eaten by the goat? Give your answer correct to 3 decimal places.

2. Larissa has a sheet of cardboard measuring 60 cm by 30 cm. She needs to cut out two identical circles, as large as possible. What is the area of cardboard remaining after the circles have been removed?

3. Casey wants to buy a tablecloth for her circular table. The table has a diameter of 2.3 m and the tablecloth is to hang 10 cm over the edge of the table. What will be the area of the tablecloth, to the nearest square metre?

4. The cross-section of a metal pipe is an annulus where the internal radius of 2 cm and the external radius is 2.3 cm. Find the area of the cross-section of the metal, correct to 2 decimal places?

5. A circle is drawn on a number plane with the centre at (3, 2). The circumference of the circle passes though the point (3, −2). What is the area of the circle, correct to 2 decimal places?

6. CHALLENGE The diagram shows a garden made from circles where the diameter AD of the large circle is 24 m, and $AB = BC = CD$. What is the area of the shaded region? Leave your answer in terms of π.

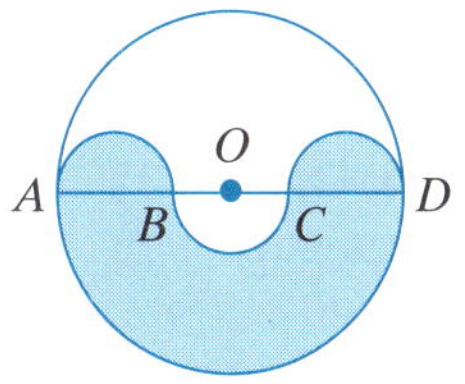

O
A B C D

Answers page 152

KEY SKILL

39 Area of circles B

HINTS

- Make sure you know the **Area of circles Hints** from Key Skill 38 on page 94.
- A semicircle is half a circle, a quadrant is one-quarter of a circle, e.g. Find the area of a semicircle with a radius of 12 cm, correct to 2 decimal places.

$$A = \frac{1}{2}\pi r^2$$
$$= \frac{1}{2} \times \pi \times 12^2$$
$$= 226.194\,6711\ \dots$$
$$= 226.19 \text{ (2 dec. pl.)}$$
$\therefore$ area is 226.19 cm^2

Reminder!

- Read the question carefully to identify what needs to be found.
- Re-read the question when you've finished your answer to make sure that the question has been answered and your solution makes sense.

Examples

The diagram shows a semicircle and a quadrant. The radius of the quadrant is 8 cm. What is the area of the shaded region? Leave your answer in terms of π.

Solution

Quadrant radius = 8 cm,
semicircle radius = 4 cm

$$A = \frac{1}{4} \times \pi \times 8^2 - \frac{1}{2} \times \pi \times 4^2$$
$$= 16\pi - 8\pi$$
$$= 8\pi$$
$\therefore$ the area is 8π cm^2.

FOCUS on ...

1. The question: Asks you to find the shaded area.
2. The information: Gives you the radii of quadrant and semicircle.
3. Your working: The semicircle is half a circle and the quadrant is one-quarter of a circle. Find the areas of both and subtract the semicircle from the quadrant:
   $A = \frac{1}{4} \times \pi \times 8^2 - \frac{1}{2} \times \pi \times 4^2$
   $= 16\pi - 8\pi$
   $= 8\pi$
4. Your answer: Make sure you write the correct units: The area is 8π cm^2.

The diagram shows three semicircles. The biggest semicircle covers parts of the two smaller semicircles. What is the area of the shaded region?

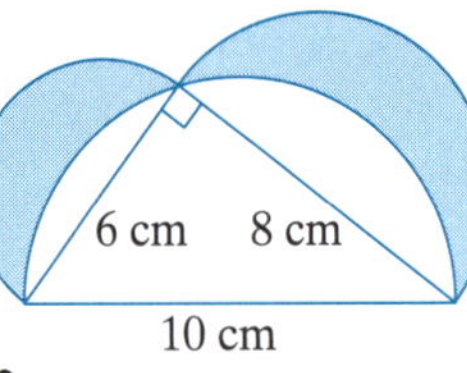

Solution

Area = Area of triangle + area of 2 semicircles − area of large semicircle

$$= \frac{1}{2} \times 8 \times 6 + \frac{1}{2} \times \pi \times 3^2 + \frac{1}{2} \times \pi \times 4^2 - \frac{1}{2} \times \pi \times 5^2$$
$$= 24$$
$\therefore$ the area is 24 cm^2.

FOCUS on ...

1. The question: Asks you to find the shaded area.
2. The information: Gives you the diameter of three semicircles.
3. Your working: Add the area of the triangle to the areas of the two smaller semicircles and then subtract the area of the larger semicircle.
   A = Area of triangle + area of 2 semicircles − area of large semicircle
   $= \frac{1}{2} \times 8 \times 6 + \frac{1}{2} \times \pi \times 3^2 + \frac{1}{2} \times \pi \times 4^2 - \frac{1}{2} \times \pi \times 5^2$
   $= 24$
4. Your answer: Make sure you write the correct units: The area is 24 cm^2.

Now try these!

1. AB is the diameter of a semicircle with centre O, and OB is the diameter of another semicircle. If $OA = 8$ cm, find the area of the shaded region, correct to 2 decimal places.

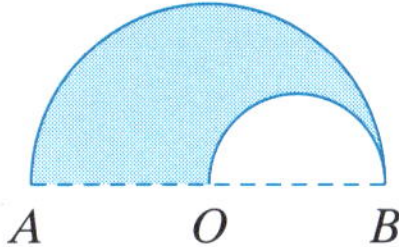

2. AOB is a quadrant with centre O. If $OA = 8$ cm and $AC = 4$ cm, find the area of the shaded region, correct to 2 decimal places.

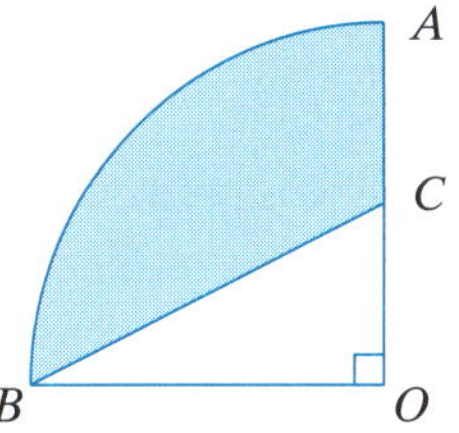

3. The diagram shows two circles with centre O. $OA = 3$ cm and $AB = 2$ cm. Find the area of the shaded region in terms of π.

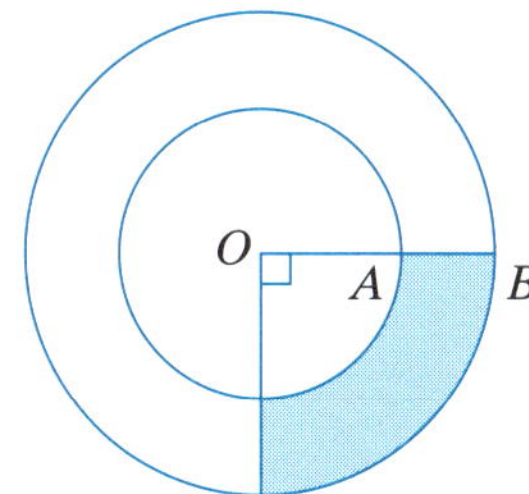

4. Two equal semicircles lie inside a square. Find the shaded area, if the perimeter of the square is 32 cm. Give your answer to 2 decimal places.

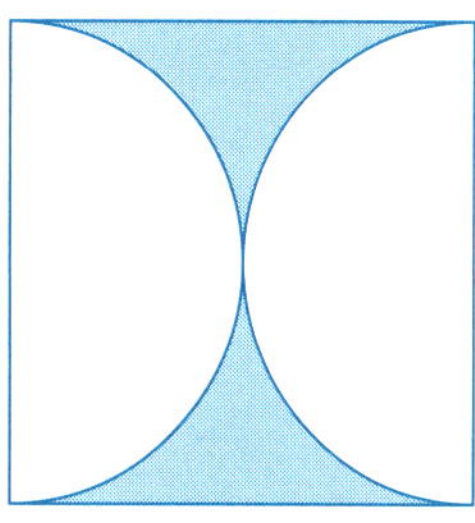

5. CHALLENGE The diagram shows a circle, centre O and equilateral triangle OPQ. M and N are midpoints of OP and OQ respectively. Find the area of the sector OMN, if $PQ = 6$ cm. Leave your answer in terms of π.

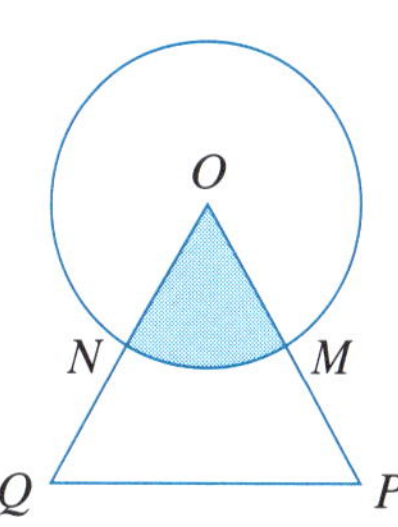

Answers page 153

REVISION TEST 11 Level of difficulty—Average

1. How many times will the tyre of a car rotate on a journey of 5000 km if the tyre has a diameter of 60 cm? Give your answer to the nearest thousand.

2. A parallelogram has a perpendicular height of 12 cm. It has the same area as a rhombus with diagonals 24 cm and 16 cm. What is the length of the base of the parallelogram?

3. A square with perimeter 48 cm has the same area as a trapezium. If the sum of the two parallel sides of the trapezium is 20 cm, find its perpendicular height.

4. Find the area of the shaded region, correct to 2 decimal places.

y
3
0
3
x

5. A running track has two straight sections of 90 m each and two semicircular sections. If the track is 400 m long, what is the radius of the semicircular sections, to the nearest metre?

6. A square has a perimeter of 40 cm. The biggest possible circle is drawn inside the square. What percentage of the square is not covered by the circle?

Answers pages 153–154

REVISION TEST 12 Level of difficulty—Challenging

1. Water is brought up from a well in a bucket. A rope attaches the bucket to a wheel of diameter 60 cm. If it takes 50 seconds to raise the bucket at a speed of 1.3 m/s, how many revolutions of the wheel are required?

2. A park has a circular pool with diameter 20 m. Around the pool is a 4-metre wide path. The path is to be resurfaced with a special coating at the rate of 1.4 m^2/L. How many litres of the coating are required?

3. The rectangle *PQRS* has a perimeter of 28 cm. The semicircle with diameter *SR* has an area of 8π cm^2. What is the perimeter of the shaded region? Leave your answer in terms of π.

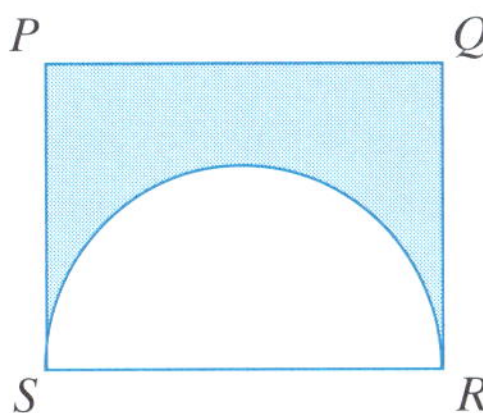

4. The shape comprises a rectangle of width 2 cm joined to a semicircle. If the area of the shape is $(6 + \frac{1}{2}\pi)$ cm^2, what is the perimeter?

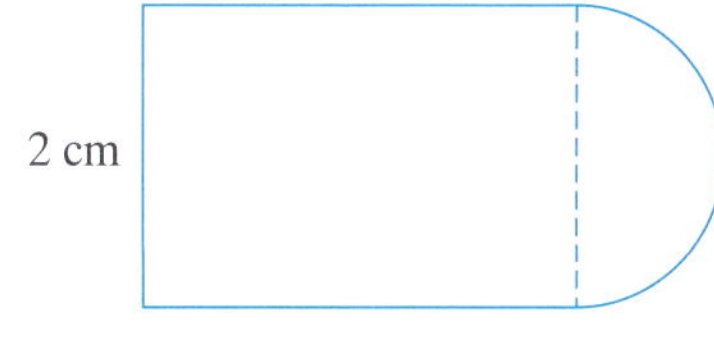

5. Every time the pedals go through a 360° rotation on a bicycle, the tyres rotate three times. If the tyres are 60 cm in diameter, what is the minimum number of complete rotations of the pedals needed for the bicycle to travel one kilometre?

6. Todd and Ken are competing in a cycling race of 20 km on a circular track of radius 60 m. They begin at the same time. Todd averages 40 km/h, while Ken averages 36 km/h. Approximately how many laps will Ken still have to complete when Todd finishes?

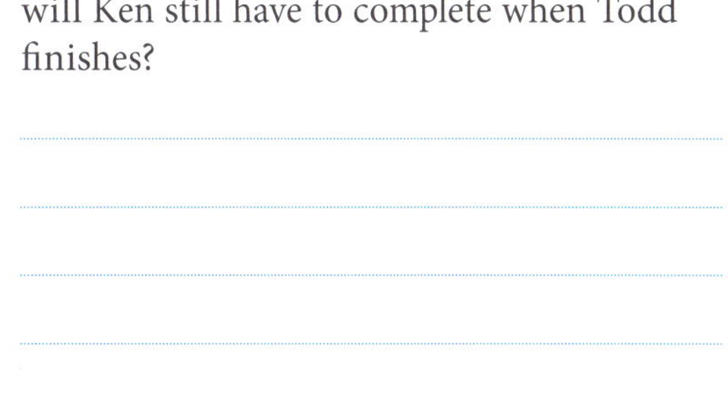

Answers pages 154–155

KEY SKILL

40 Volume of prisms

HINTS

- Volume is a measure of the space inside a solid shape.
- The formulae are:
  Prism: $V = Ah$
  Rectangular prism: $V = lbh$
  Cube: $V = s^3$

Reminder!

- Read the question carefully to identify what needs to be found.
- Re-read the question when you've finished your answer to make sure that the question has been answered and your solution makes sense.

Examples

A cube has the same volume as a rectangular prism with dimensions 18 cm, 12 cm and 8 cm. Find the length of a side of the cube.

Solution

Rectangular prism: $V = lbh$
$= 18 \times 12 \times 8$
$= 1728$

$\therefore$ the volume is 1728 cm^3.

Cube: $V = s^3$
$s^3 = 1728$
$s = \sqrt[3]{1728}$
$= 12$

$\therefore$ the side length is 12 cm.

FOCUS on ...

1. **The question:** Asks you to find the length of the side of a cube with identical volume as a prism.
2. **The information:** Gives you the dimensions of a rectangular prism.
3. **Your working:** Use the formula for the volume of a rectangular prism.
   $V = lbh$
   $= 18 \times 12 \times 8$
   $= 1728$
   $\therefore$ the volume is 1728 cm^3.
   Use the formula for the volume of a cube.
   $V = s^3$
   $s^3 = 1728$
   $s = \sqrt[3]{1728}$
   $= 12$
4. **Your answer:** Make sure you write the correct units:
   The side length is 12 cm.

The volume of a rectangular prism is the same as a cube with side length 24 cm. If the length and width of the prism are both 16 cm, what is the height of the prism?

Solution

Cube: $V = s^3$
$= 24^3$
$= 13\,824$

$\therefore$ the volume is 13 824 cm^3

Rectangular prism: $V = lbh$
$13\,824 = 16 \times 16 \times h$
$256h = 13\,824$
$h = 13\,824 \div 256$
$= 54$

$\therefore$ the height of the prism is 54 cm.

FOCUS on ...

1. **The question:** Asks you to find the height of the prism with identical volume as a cube.
2. **The information:** Gives you the dimensions of a cube and the length and width of the prism.
3. **Your working:** Use the formula for the volume of a cube.
   $V = s^3$
   $= 24^3$
   $= 13\,824$
   $\therefore$ the volume is 13 824 cm^3.
   Use the formula for the volume of a rectangular prism.
   $V = lbh$
   $13\,824 = 16 \times 16 \times h$
   $13\,824 = 256h$
   $h = 13\,824 \div 256$
   $= 54$
4. **Your answer:** Make sure you write the correct units:
   The height of the prism is 54 cm.

Now try these!

1. A cube and a rectangular prism have the same volume. The dimensions of the rectangular prism are 52 cm, 26 cm and 13 cm. What is the length of the sides of the cube?

2. A rectangular prism has a length of 24 cm and a breadth of 18 cm. If the volume is 6912 cm^3, what is the height of the prism?

3. The volume of a cube is the same as the volume of a rectangular prism with length 18 cm and breadth 14 cm. If the cube has a side length of 21 cm, what is the height of the rectangular prism?

4. After an afternoon storm Farmer Brown measures 20 mm in his rain gauge. If the rainfall was consistent across his whole farm of 27 hectares, what volume of rain fell on his farm? Give your answer in cubic metres.

5. The cross-section of a 3-mm thick medallion is in the shape of a trapezium. The height of the trapezium is 1.5 cm and the parallel sides are 1.1 cm and 1.8 cm. What is the volume of the medallion?

6. CHALLENGE Chocolate bars are produced in the shape of a triangular prism. The base of the bar is a right-angled isosceles triangle with two sides 30 mm. Chocolate to produce the bars is melted from a block in the shape of a square-based prism with dimensions 20 cm, 20 cm and 9 cm. If chocolate from the block can make 400 chocolate bars, what is the depth of each bar?

Answers page 155

KEY SKILL

41 Capacity

HINTS

- Capacity is the amount of liquid or pourable substance that a solid can hold.
- Make sure you know the **Volume of prisms Hints** from Key Skill 40 on page 100.
- To convert between volume and capacity we use:
  - 1000 cm^3 = 1000 mL = 1 L
  - 1 m^3 = 1000 L = 1 kL

  e.g. Find the capacity of a cube with sides 8 cm.

  $$\text{Volume} = 8^3 = 512$$

  ∴ 512 cm^3

  Capacity = 512 mL

Reminder!

- Read the question carefully to identify what needs to be found.
- Re-read the question when you've finished your answer to make sure that the question has been answered and your solution makes sense.

Examples

A container is in the shape of a rectangular prism with a length of 25 cm and a width of 16 cm. It contains 2 litres of water. What is the height of water in the container?

Solution

2 litres = 2000 mL = 2000 cm^3

∴ the volume is 2000 cm^3

Rectangular prism: $V = lbh$

$$2000 = 25 \times 16 \times h$$
$$400h = 2000$$
$$h = 2000 \div 400$$
$$= 5$$

∴ the water is 5 cm deep.

FOCUS on ...

1. The question: Asks you to find the height (or depth) of the water in a rectangular prism.
2. The information: Gives you the length and width of the water in the prism and the amount of water in the prism.
3. Your working: You need to change the amount of water into cm^3 by using 1 L = 1000 cm^3.
   2 litres = 2000 mL = 2000 cm^3
   ∴ the volume is 2000 cm^3.
   Use the formula for the volume of a rectangular prism. $V = lbh$
   $$2000 = 25 \times 16 \times h$$
   $$400h = 2000$$
   $$h = 2000 \div 400 = 5$$
4. Your answer: Make sure you write the correct units: The water is 5 cm deep.

An empty water tank is in the shape of a prism with a base area of 1600 cm^2. Water from a tap flows into the tank at a constant rate of 8 litres per minute. What is the height of water in the tank after 10 minutes?

Solution

$$\text{Amount of water} = 8 \times 10 = 80$$

∴ the amount of water is 80 litres, or 80 000 cm^3.

$$\text{Volume} = Ah$$
$$80\,000 = 1600 \times h$$
$$h = 80\,000 \div 1600$$
$$= 50$$

∴ the depth of water is 50 cm.

FOCUS on ...

1. The question: Asks you to find the height (or depth) of the water in a prism.
2. The information: Gives you the area of the base, the rate of water flow into the tank and the time of flow.
3. Your working: You need to multiply to find the amount of water:
   $$\text{Amount of water} = 8 \times 10 = 80$$
   ∴ the amount of water is 80 litres.
   You need to change the amount of water into cm^3 by using 1L = 1000 cm^3:
   80 litres = 80 000 mL = 80 000 cm^3
   Use the formula for the volume of a rectangular prism: $V = Ah$
   $$80\,000 = 1600 \times h$$
   $$h = 80\,000 \div 1600 = 50$$
4. Your answer: Make sure you write the correct units: The water is 50 cm deep.

Now try these!

1. A fish tank is in the shape of a rectangular prism and has a capacity of 72 litres. The length is 60 cm and the breadth is 40 cm. What is the height of the tank?

2. A container has a cross-sectional area of 480 cm^2. Water is being poured into a container at a constant rate of 80 mL/second. How long will it take for the water to be 12 cm deep?

3. A container in the shape of a cube is three-quarters full of water. Lana fills the container by adding 54 mL of water. What is the length of each side of the container?

4. A container is in the shape of a rectangular prism with base 20 cm by 16 cm and a height of 12 cm. Water has been poured into the container to a depth of 5 cm. How much more water is required to completely fill the container?

5. An 18-metre length of aluminium gutter in the shape of a trapezoidal prism is placed on the front of a house. The two parallel sides of the trapezium are 12 cm and 15 cm. How deep is the gutter if it holds 388.8 litres of water?

6. CHALLENGE A drink dispenser is in the shape of a rectangular prism. It has a base area of 500 cm^2 and a height of 20 cm. Chris mixes a cordial of concentrate and water in the ratio of 1 : 4 and adds it to the empty dispenser. If he uses 6.4 L of water, what percentage of the dispenser will be filled?

Answers pages 155–156

KEY SKILL

42 Volume of cylinders

HINTS

- Volume is a measure of the space inside a solid shape.
- The formula for volume of a cylinder is: $V = \pi r^2 h$.

Reminder!

- Read the question carefully to identify what needs to be found.
- Re-read the question when you've finished your answer to make sure that the question has been answered and your solution makes sense.

Examples

A can of beans has a height of 10 cm and a diameter of 7 cm. Find the volume of the can, to the nearest centimetre.

Solution

$$\begin{aligned} V &= \pi r^2 h \\ &= \pi \times 3.5^2 \times 10 \\ &= 384.845\,1001\ldots \\ &= 385 \text{ (nearest whole)} \end{aligned}$$

$\therefore$ the volume is 385 cm^3.

FOCUS on ...

1. The question: Asks you to find the volume of the cylinder.
2. The information: Gives you the diameter and the height of the cylinder.
3. Your working: Find the radius of the circle.
   As $7 \div 2 = 3.5$, the radius is 3.5 cm.
   Use the formula for the volume of the cylinder.
   $$\begin{aligned} V &= \pi r^2 h \\ &= \pi \times 3.5^2 \times 10 \\ &= 384.845\,1001\ldots \end{aligned}$$
   Correct your answer to nearest whole.
   $$= 385 \text{ (nearest whole)}$$
4. Your answer: Make sure you write the correct units:
   The volume is 385 cm^3.

The height of a cylinder is 12 cm and its volume is 192π cm^3. What is the radius of the cylinder?

Solution

$$\begin{aligned} V &= \pi r^2 h \\ 192\pi &= \pi \times r^2 \times 12 \\ 192\pi &= 12\pi r^2 \\ r^2 &= 192\pi \div 12\pi \\ &= 16 \\ r &= 4 \;\; (r > 0) \end{aligned}$$

$\therefore$ the radius is 4 cm.

FOCUS on ...

1. The question: Asks you to find the radius of the cylinder.
2. The information: Gives you the height and the volume of the cylinder.
3. Your working: Use the formula for the volume of the cylinder.
   $$\begin{aligned} V &= \pi r^2 h \\ 192\pi &= \pi \times r^2 \times 12 \\ &= 12\pi r^2 \\ r^2 &= 192\pi \div 12\pi \\ &= 16 \\ r &= 4 \;\; (r > 0) \end{aligned}$$
4. Your answer: Make sure you write the correct units:
   The radius is 4 cm.

Now try these!

1. Two cylinders are being compared. Cylinder A has a diameter of 12 cm and a height of 16 cm. The diameter of Cylinder B is 2 cm smaller than the other cylinder but it is 4 cm taller. Which cylinder has the greater volume and by how much? Give your answer to the nearest cubic centimetre.

2. Grace placed a can under a dripping tap. The can is in the shape of a cylinder with a diameter of 12 cm and a height of 10 cm. The tap is dripping at the rate of 1 mL/second. How long, to the nearest minute, would it take for the can to fill?

3. Connor wants to construct a cylindrical container to hold enough water for his pet fish. He has learned that his fish needs 20 litres of water. If he constructs his tank with a diameter of 24 cm, how tall must he make the tank to hold the right amount of water? Give your answer to the nearest centimetre.

4. The capacity of a cylinder is 1.6 litres. If the height is 12 cm, what is the diameter of the cylinder? Give your answer correct to 2 decimal places.

5. A family purchased a circular swimming pool which has a diameter of 3.2 m and wall height of 1.2 m. The manufacturer recommends that the pool only be filled to 75% of its capacity. How much water (in litres) should be used to fill it to the recommended level?

6. CHALLENGE A hollow metal pipe is 45 cm in length and has an external diameter of 10 cm. If the thickness of the pipe is 0.5 cm and the metal has a mass of 11 grams/cm^3, what is the total mass of the pipe? Give your answer to the nearest kilogram.

Answers page 156

KEY SKILL

43 Pythagoras' theorem A

HINTS

- In a right-angled triangle the square of the hypotenuse equals the sum of the squares of the other two sides:

$c^2 = a^2 + b^2$

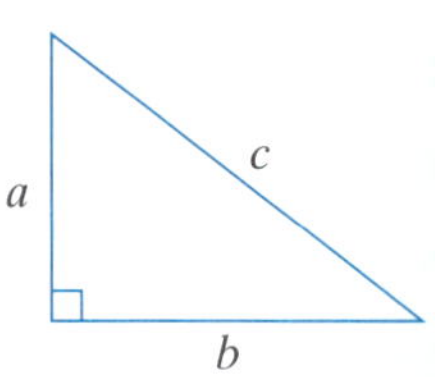

Reminder!

- Read the question carefully to identify what needs to be found.
- Re-read the question when you've finished your answer to make sure that the question has been answered and your solution makes sense.

Examples

A plane leaves Whyalla airport and flies 320 km north and then turns and flies 210 km east. How far is the plane from Whyalla, to the nearest kilometre?

Solution

$$x^2 = 320^2 + 210^2$$
$$= 146\,500$$
$$x = \sqrt{146\,500}$$
$$= 382.753\,1842\ldots$$
$$= 383 \text{ (nearest whole)}$$

$\therefore$ the distance is 383 km.

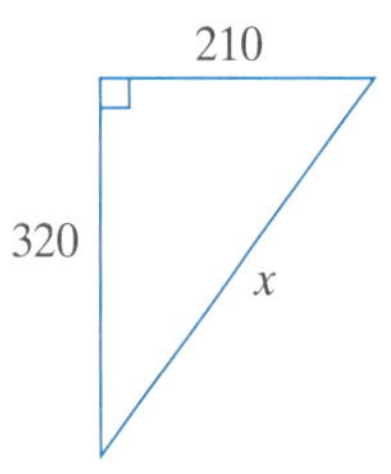

FOCUS on ...

1. **The question:** Asks you to find the distance.
2. **The information:** Gives you the distances travelled north and east.
3. **Your working:** Draw a diagram.
   Let the distance = x, and use Pythagoras' theorem.
   $$x^2 = 320^2 + 210^2$$
   $$= 146\,500$$
   $$x = \sqrt{146\,500}$$
   $$= 382.753\,1842\ldots$$
   $$= 383 \text{ (nearest whole)}$$
4. **Your answer:** Make sure you write the correct units:
   The distance is 383 km.

A 3.6-metre ladder is leaning on a wall. The base of the ladder is 1.1 m from the wall. How far up the wall is the top of the ladder, correct to 2 decimal places?

Solution

$$x^2 = 3.6^2 - 1.1^2$$
$$= 11.75$$
$$x = \sqrt{11.75}$$
$$= 3.427\,8273\ldots$$
$$= 3.43 \text{ (2 dec. pl.)}$$

$\therefore$ the height is 3.43 m.

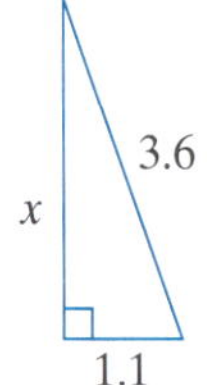

FOCUS on ...

1. **The question:** Asks you to find the height of the ladder.
2. **The information:** Gives you the length of the ladder and the distance of the base of the ladder from the wall.
3. **Your working:** Draw a diagram.
   Let the height of the ladder = x, and use Pythagoras' theorem.
   $$x^2 = 3.6^2 - 1.1^2$$
   $$= 12.96 - 1.21$$
   $$= 11.75$$
   $$x = \sqrt{11.75}$$
   $$= 3.427\,8273\ldots$$
   Correct your answer to 2 decimal places.
   $$= 3.43$$
4. **Your answer:** Make sure you write the correct units:
   The height is 3.43 m.

Now try these!

1. An orientation course has runners leaving the starting point and travelling 4 km in a southerly direction before turning and running 2.8 km in a westerly direction. They then run directly back to the starting point. How far is the third leg, to the nearest kilometre?

2. A pole is 6.4 m high and is supported by an 8-metre length of wire attached to the top of the pole. The wire is anchored in a concrete base. How far is the base from the bottom of the pole?

3. A can of drink is in the shape of a cylinder with radius 4 cm and height 15 cm. What is the length of the longest straw that will fit inside the can?

4. A park is in the shape of a square. Andrew walked diagonally across the park while Rose walked along two of the sides. Express the distance that Andrew walked as a percentage of Rose's distance. Give your answer to the nearest percentage.

5. A rhombus has an area of 96 cm^2. If one diagonal is 12 cm, what is the perimeter of the rhombus?

6. CHALLENGE When Lachlan drives to work he picks up his workmate Ryan. Lachlan lives directly 18 km south of Ryan and their workplace is 15 km directly east of Ryan's house. Lachlan always drives at 40 km/h. How much time would Lachlan save if he did not have to pick up Ryan?

Answers pages 156–157

KEY SKILL

44 Pythagoras' theorem B

HINTS

- Make sure you know the **Pythagoras' theorem Hints** from Key Skill 43 on page 106.

Reminder!

- Read the question carefully to identify what needs to be found.
- Re-read the question when you've finished your answer to make sure that the question has been answered and your solution makes sense.

Examples

A paddock is in the shape of a right-angled triangle with the longest side being 18 m. If one of the other sides is 10 m, what is the cost of fencing the paddock if materials and labour cost \$2.35/m? Answer to the nearest dollar.

Solution

$$
\begin{aligned}
x^2 &= 18^2 - 10^2 \\
&= 224 \\
x &= 14.966\,629\,55\ldots \\
&= 14.97 \text{ (2 dec. pl.)} \\
\text{Perimeter} &= 18 + 10 + 14.97 \\
&= 42.97 \\
\text{Cost} &= 2.35 \times 42.97 \\
&= 100.9795 \\
&= 101 \text{ (nearest whole)}
\end{aligned}
$$

$\therefore$ cost is \$101.

FOCUS on...

1. The question: Asks you to find the cost of fencing.
2. The information: Gives you two sides of the triangle and the cost of fencing per metre.
3. Your working: Apply Pythagoras' theorem when unknown not on hypotenuse.
   $x^2 = 18^2 - 10^2$
   $= 224$
   $= \sqrt{224}$
   $x = 14.966\,629\,55\ldots$
   Correct your answer to 2 decimal places.
   $= 14.97$ (2 dec. pl.)
   Add the sides to find the perimeter.
   $P = 18 + 10 + 14.97$
   $= 42.97$
   Multiply the cost per metre by the length.
   $C = 2.35 \times 42.97$
   $= 100.9795$ Correct your answer to the nearest whole.
   $= 101$ (nearest whole)
4. Your answer: Make sure you write the correct units:
   The cost is \$101.

A square sheet of plywood has a diagonal of 2.8 m. What is the cost of covering the sheet with material that costs \$8.35/m^2?

Solution

$$
\begin{aligned}
x^2 + x^2 &= 2.8^2 \\
2x^2 &= 7.84 \\
x^2 &= 3.92 \\
\text{Area} &= x \times x \\
&= x^2 \\
&= 3.92 \\
\text{Cost} &= 8.35 \times 3.92 \\
&= 32.732 \\
&= 32.73 \text{ (2 dec. pl.)}
\end{aligned}
$$

$\therefore$ cost is \$32.73.

FOCUS on...

1. The question: Asks you to find the cost of covering the sheet of plywood.
2. The information: Gives you the sides of the square and the cost of covering per square metre.
3. Your working: Apply Pythagoras' theorem.
   $x^2 + x^2 = 2.8^2$
   $2x^2 = 7.84$
   $x^2 = 3.92$
   $\text{Area} = x \times x$
   $= x^2$
   $= 3.92$
   $\text{Cost} = 8.35 \times 3.92$
   $= 32.732$
   Correct your answer to 2 decimal places.
   $= 32.73$
4. Your answer: Make sure you write the correct units:
   The cost is \$32.73.

Now try these!

1. A right-angled paddock has an area of 80 m^2. The base of the triangle is 16 m. The paddock is to be fenced at a cost of \$3.45/m. Find the cost, to the nearest dollar.

2. A square garden has a diagonal of 32 m. What is the cost of fertilising the garden at the rate of $\$0.04/\text{m}^2$?

3. A circle is drawn through the vertices of a square *ABCD*. If $AB = 4$ cm, find the area of the shaded region, in terms of π.

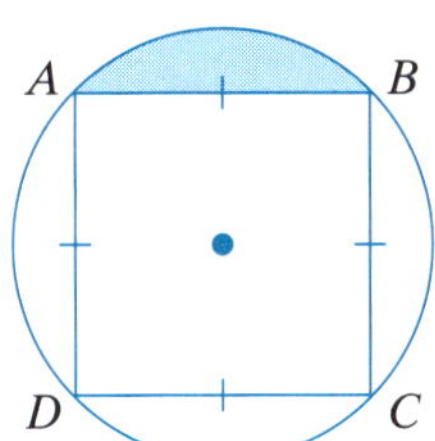

4. A ladder is 2.5 m long and leans against a wall. The end of the ladder is 70 cm from the base of the wall. If the top of the ladder slips 40 cm down the wall, how far does the end of the ladder move out from the wall?

5. Two yachts are sailing away from the same buoy. Yacht A is sailing east at 12 km/h and Yacht B is sailing south at 16 km/h. How far apart are the yachts after two and a half hours?

6. CHALLENGE A square *OEDC* sits inside a quadrant *OADB*. If $DC = 6$ cm, find the area of the shaded region, leaving your answer in terms of π.

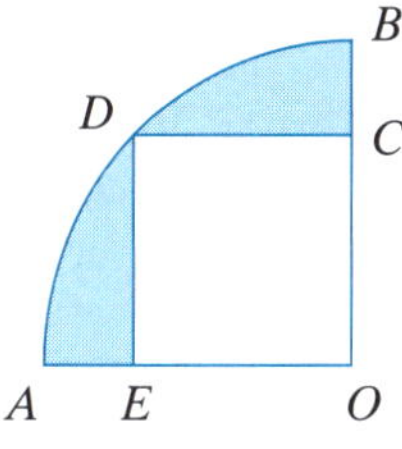

Answers pages 157–158

KEY SKILL

45 Time

HINTS

- Calculators with a 'DMS' button (for degrees, minute, seconds) are useful for time calculations,

  e.g. What is the time difference between 4 h 39 min 33 s and 2 h 47 min 58 s?

  4 DMS 39 DMS 33 − 2 DMS 47 DMS 58 = 1 h 51 min 35 s
- Locations across Australia (or the world) have different time zones. The time difference is identified in the question,

  e.g. In winter, Perth is 2 hours behind Sydney. If it is 9 am in Sydney, what is the local time in Perth?

  As 9 − 2 = 7, then it is 7 am in Perth.

Reminder!

- Read the question carefully to identify what needs to be found.
- Re-read the question when you've finished your answer to make sure that the question has been answered and your solution makes sense.

Examples

The table shows the times in different cities in December in 24-hour time.

City	Time
Adelaide	0750
Auckland	1020
Brisbane	0720
Perth	0520
Sydney	0820

If it is 1835 in Perth, what is the local time in Auckland, expressed in 12-hour time?

Solution

The time in Auckland is 5 h ahead of Perth.
1835 pm plus 5 h is 2335.
As 2335 = 11:35 pm, it is 11:35 pm in Auckland.

FOCUS on ...

1. **The question:** Asks you to find the local time in Auckland.
2. **The information:** Gives you the local time in Perth and the time difference between the two cities.
3. **Your working:** Use the table to calculate the time difference between Auckland and Perth.
   1020 − 0520 = 5 h
   Auckland is 5 h ahead of Perth.
   Add 5 h on to 1835.
   1835 plus 5 h = 2335
   Convert to 24-hour time.
   2335 = 11:35 pm
4. **Your answer:** Make sure you write the correct units:
   The time in Auckland is 11:35 pm.

James is to fly from Brisbane to Adelaide on Christmas Eve. He leaves Brisbane at 1025 and the flight takes 2 h 20 min. What time is it in Adelaide when he arrives?

Solution

As 1025 plus 30 min is 1055, the plane takes off at 1055 Adelaide time. From 1055 plus 2 h is 1255, plus 5 min is 1300, plus 15 min is 1315. The plane arrives in Adelaide at 1315.

FOCUS on ...

1. **The question:** Asks you to find the local time in Adelaide when James arrives.
2. **The information:** Gives you the time in Brisbane when the flight leaves and the duration of the flight. The table is used to find the time difference.
3. **Your working:** Use the table to calculate the time difference between Brisbane and Adelaide.
   0750 − 0720 = 30 min. Adelaide is 30 min ahead.
   You find the local time in Adelaide when the flight takes off: 1025 plus 30 min is 1055.
   Add the flight time to Adelaide's local time.
   1055 plus 2 h 20 min = 1315
4. **Your answer:** Make sure you write the correct units:
   The local time in Adelaide is 1315.

Now try these!

The table shows the local times in different cities in mid-March and is used to answer questions 1 to 6.

City	Time
Auckland	0830 Saturday
Perth	0330 Saturday
Moscow	2330 Friday
Rome	2030 Friday
San Francisco	1230 Friday
Honolulu	0930 Friday

1. If it is 1450 Monday in Perth, what is the local time in San Francisco, expressed in 12-hour format?

2. Anais leaves Rome at 4:50 pm on Tuesday and flies to Perth. The flight takes 18 hours 20 minutes. What is the local time when she arrives in Perth?

3. Sydney is 2 hours behind Auckland. At 4:30 pm on Sunday Mim is in Sydney and chats to her mother in San Francisco for 40 minutes. What time is it in San Francisco when they finish chatting?

4. Chicago is 2 hours ahead of San Francisco. If it is 6:30 pm Thursday in Chicago, what is the local time in Auckland?

5. New York is 3 hours ahead of San Francisco. Phoebe left New York at 8:40 am and flew to Honolulu. The flight took 10 hours 15 minutes. What was the local time in Honolulu when she arrived?

6. CHALLENGE A news conference in London commenced at 9:45 on Tuesday morning and concluded 40 minutes later. Amber watched the live news conference in her home in Adelaide. Adelaide is 2 hours 30 minutes ahead of Perth and London is 1 hour behind Rome. What was the local time in Adelaide when the news conference finished?

Answers page 158

REVISION TEST 13 Level of difficulty—Average

1. The capacity of a cube is the same as the capacity of a rectangular prism with length 36 cm and breadth 20 cm. If the cube has a side length of 24 cm, what is the height of the rectangular prism?

2. Aleisha's cylindrical water tank has a capacity of 8 kilolitres. The tank has a diameter of 1.8 m. When the tank is half-full what is the height of the water? Give your answer to the nearest centimetre.

3. Cheryl has a box in the shape of a rectangular prism with dimensions 28 cm by 16 cm by 10 cm. She has some cubes with sides 4 cm to place in the box. What is the maximum number of cubes that can fit into the box?

4. A cube with side 10 cm is filled with water. The water is then poured into a cylinder with a radius of 10 cm. What is the height of water in the cylinder, in terms of π?

5. The width of a rectangular paddock is 48 m. If the diagonal of the paddock is 73 m, what is the perimeter of the paddock?

6. Hui lives in Xi'an and his cousin Li lives in Dunedin. When it is 10:30 am in Xi'an it is 2:30 pm in Dunedin. The boys start playing an online game at 4:50 pm Xi'an time. If they finish playing 90 minutes later, what is the local time in Dunedin?

Answers pages 158–159

REVISION TEST 14 Level of difficulty—Challenging

1 ABO is a quadrant and $CDEO$ is a rectangle with sides 8 cm and 6 cm. What is the length of arc AB, in terms of π?

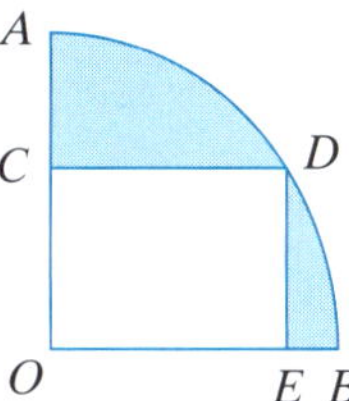

2 Each morning to get to work, Tom leaves home and drives 8 km west, 10 km north, 4 km west and 6 km north. What is the straight-line distance from home to work?

3 At 7 am, Mei turned on a hose to fill an empty rectangular container at the rate of 8 litres per minute. At the same time water began leaking out of the container's faulty tap at the rate of 0.4 litres per minute. Mei turned off the hose when the container was full. If the container's dimensions were 80 cm by 70 cm by 38 cm, when was the container empty again?

4 The table shows the time difference between two cities.

City	Time
Brisbane	0400 Wednesday
Dubai	2200 Tuesday

Sophie left Brisbane at 9:00 pm on Friday and landed in Dubai. The plane flew 11 890 km and averaged a speed of 820 km/h. What was the local time in Dubai when she arrived?

5 A cylinder with diameter 12 cm and height 16 cm is half-filled with water. The water is then poured into another cylinder with a radius of 8 cm. What is the depth of the water in the second container?

6 The diagram shows a cube $ABCDEFGH$ of side 6 cm. M is the midpoint of the side BF. What is the shortest distance from D to M?

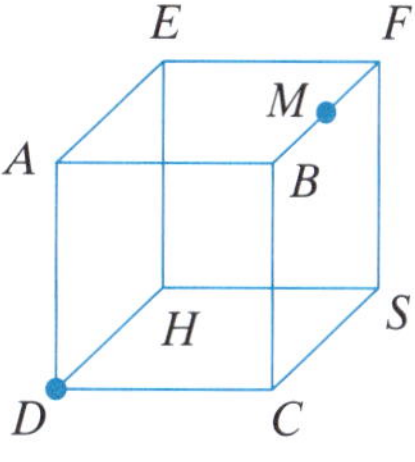

Answers page 159

KEY SKILL

STATISTICS AND PROBABILITY

46 Statistics A

HINTS

- Mean is the average. You add the scores and divide by the number of scores.

Reminder!

- Read the question carefully to identify what needs to be found.
- Re-read the question when you've finished your answer to make sure that the question has been answered and your solution makes sense.

Examples

The mean mass of 4 boys is 67 kg. Another two boys have a mean mass of 70 kg. What is the mean mass of the 6 boys?

Solution

Mean mass of 6 boys

$= (67 \times 4 + 70 \times 2) \div 6$

$= 68$

$\therefore$ the mean mass is 68 kg.

FOCUS on...

1. **The question:** Asks you to find the mean mass of the 6 boys.
2. **The information:** Gives you the mean mass of 4 boys and the mean mass of another 2 boys.
3. **Your working:** Multiply to find the total mass of 4 boys.
   Total mass of 4 boys $= 67 \times 4$
   $= 268$
   Multiply to find total mass of 2 boys.
   Total mass of 2 boys $= 70 \times 2$
   $= 140$
   Add to find total mass of the 6 boys.
   Total mass of 6 boys $= 268 + 140$
   $= 408$
   Divide to find the mean mass of the 6 boys.
   Mean mass of 6 boys $= 408 \div 6$
   $= 68$
4. **Your answer:** Make sure you write the correct units:
   The mean mass is 68 kg.

The mean of the results of 5 tests is 70. What must Mike score in the sixth test to have an overall mean of 75?

Solution

Mean of 5 tests $= 70$

Total of 5 tests $= 70 \times 5$

$= 350$

Mean of 6 tests $= 75$

Total of 6 tests $= 75 \times 6$

$= 450$

Difference $= 450 - 350$

$= 100$

$\therefore$ Mike needs to score 100 in the sixth test.

FOCUS on...

1. **The question:** Asks you to find Mike's result in the sixth test.
2. **The information:** Gives you the mean of the 5 tests and the required overall mean.
3. **Your working:** Multiply to find the total of 5 tests.
   Total of 5 tests $= 70 \times 5$
   $= 350$
   Multiply to find the total of 6 tests.
   Total of 6 tests $= 75 \times 6$
   $= 450$
   Subtract the two totals.
   Difference $= 450 - 350$
   $= 100$
4. **Your answer:** Make sure you write the correct units:
   Mike needs 100 in the sixth test.

Now try these!

1. A small engineering company has 6 employees. The mean salary each week is $1295. If the salaries of 5 of the employees are $1160, $1210, $1290, $1300 and $1340, what is the salary of the sixth employee?

2. In 8 basketball games, Jerome scored a mean of 12 points. After his next game his mean had dropped by one. How many points did he score in the ninth game?

3. Xui Li received scores of 85, 82, 91 and 80 on her first 4 science assignments of the year. If her goal is to have a mean of 85 for her first 5 assignments, what must she score in her fifth assignment to achieve this goal?

4. In 4 tests out of 100, Connor averaged 83%. What was the lowest possible result he could have scored in any one test?

5. In a term, Belinda has to complete 5 fortnightly tests. She needs to average at least 74 for the term. Her first 4 test results were 73, 63, 66 and 71. What is the lowest score she can receive in her final test?

6. CHALLENGE In the first 8 games of her basketball season, Avery scored a mean of 11.5 points. In the ninth game she only scored 4 points. What will she need to score in the tenth game to have an overall mean of 11 points per game?

Answers page 160

KEY SKILL

47 Statistics B

HINTS

- Mean is the average. You add the scores and divide by the number of scores.
- Median is the middle score.
- Mode is the most common score.
- Range is the highest score minus the lowest score.

Reminder!

- Read the question carefully to identify what needs to be found.
- Re-read the question when you've finished your answer to make sure that the question has been answered and your solution makes sense.

Examples

Six students scored the following results in a quiz: 4, 2, 8, 8, x, 5. If the median is the same as the range, what is the value of x?

Solution

Range $= 8 - 2$
$= 6$

If the median is 6, then the scores are

2, 4, 5, x, 8, 8

If the median is 6 then $x = 7$.

$\therefore$ x must be 7.

FOCUS on...

1. The question: Asks you to find the value of x.
2. The information: Gives you 5 known scores and one score called x. Also, the range = median.
3. Your working: Find the range by finding the difference between the largest and smallest scores: $8 - 2 = 6$.
   Therefore the median is 6. Arrange the scores in ascending order. As the range is 6, then x must be between 2 and 8. Also, there is an even number of scores.
   2, 4, 5, x, 8, 8.
   $\therefore$ x is 7.
4. Your answer: The value of x is 7.

Simone used the frequency table to display the results of a test of the 15 students in her class.

Score	Frequency
b	6
c	a
d	5

It is known that the range is 5, the mode is 12 and the median is 14. What are the values of a, b, c and d, if $b < c < d$?

Solution

If the total is 15, then $a = 4$.

This means the mode $= b = 12$. If the range is 5, then the highest score $= d = 17$. As there are 15 scores, then the middle score is the eighth score. This means $c = 14$.

$\therefore$ $a = 4$, $b = 12$, $c = 14$, $d = 17$.

FOCUS on...

1. The question: Asks you to find the values of a, b, c and d.
2. The information: Gives you the total number of scores, a frequency table with some data and the value of the range, mode and median.
3. Your working: To find the value of a subtract the total of the known frequencies from 15. This means $a = 4$. Comparing the three frequencies (6, 4, 5), the highest frequency is 6. This means $b = 12$. As the range is 5 then $12 + 5 = 17$ is the largest score. This means $d = 17$.
   Finally, as there were 15 scores, the middle score is the eighth score. This means c is the median and $c = 14$.
4. Your answer: $a = 4$, $b = 12$, $c = 14$, $d = 17$.

Now try these!

1. Six students scored the following results in a quiz: 3, 9, 5, x, 6, 10. If the median is the same as the range, what is a possible value of x?

2. The frequency table records the number of children living in the homes of students in a Year 8 class of 30 students.

Score	Frequency
b	4
c	a
d	9
4	7

The range is the same as the median, and the mode is 2. If $b < c < d$, what are the values of a, b, c and d?

3. The mean of 8 scores is 10 and the median is 11. If the scores in ascending order are 3, 5, 7, 10, x, 13, y, 17, find the values of x and y.

4. A hockey team recorded the number of goals scored in 6 games. The team scored at least 2 goals in every game and scored 4 goals in exactly 2 games. If the mode, median, mean and range have the same value, how many goals were scored in each of the other games?

5. The table shows the frequency of scores.

Score	b	c	d
Frequency	2	6	a

It is known that the set of 15 scores has a range of 5, a mode of 8 and a median of 5. If $b < c < d$, find the values of a, b, c and d.

6. CHALLENGE Kelvin could not remember his scores from 5 history tests marked out of 100. He did remember that the mean was 80, the median was 83, and the mode was 84. If all his scores were whole numbers, what was the lowest score he could have received on any one test?

Answers pages 160–161

KEY SKILL

48 Probability A

HINTS

- The probability of an event is a number from 0 to 1 indicating the chance that the event will occur.
- The probability of an event occurring is given as

$$P(E) = \frac{\text{number of favourable outcomes}}{\text{total number of outcomes}}$$

e.g. A die is rolled. What is the probability of rolling a 5?

$$P(5) = \frac{1}{6}$$

Reminder!

- Read the question carefully to identify what needs to be found.
- Re-read the question when you've finished your answer to make sure that the question has been answered and your solution makes sense.

Examples

A bag contains 3 red balls and 7 blue balls. A ball is chosen at random. What is the probability that the ball is blue?

Solution

Total of 10 balls in the bag

$$P(\text{blue}) = \frac{7}{10}$$

$\therefore$ the probability of selecting a blue ball is $\frac{7}{10}$.

FOCUS on ...

1. **The question:** Asks you to find the chance that a blue ball is selected at random.
2. **The information:** Gives you the numbers of red and blue balls in a bag.
3. **Your working:** As 3 + 7 = 10, then there are 10 balls in the bag and 7 of those 10 balls are blue. This means the chance or probability of choosing a blue ball is 7 out of 10, or $\frac{7}{10}$.
4. **Your answer:** The probability is $\frac{7}{10}$.

The letters in the word PROBLEMS are written on identical cards and placed in a box. A card is chosen at random. What is the probability that the letter is a consonant?

Solution

There are 8 letters, and the consonants are P, R, B, L, M, S.

$$P(\text{consonant}) = \frac{6}{8} = \frac{3}{4}$$

$\therefore$ the probability of selecting a consonant is $\frac{3}{4}$.

FOCUS on ...

1. **The question:** Asks you to find the chance that a consonant is chosen at random.
2. **The information:** Gives you the letters of the word PROBLEMS.
3. **Your working:** There are 8 letters in the word PROBLEMS and 6 of those 8 letters are consonants. This means the chance or probability of choosing a consonant is 6 out of 8, or $\frac{6}{8} = \frac{3}{4}$.
4. **Your answer:** The probability is $\frac{3}{4}$.

Now try these!

1. A bag contains 6 yellow and 4 red balls. If a ball is chosen at random, what is the probability that it is yellow?

2. Tickets numbered 1 to 20 are placed in a bag and mixed up. A ticket is drawn at random. What is the probability that the number is a multiple of 2 or 3?

3. A die has been relabelled using 1, 2, 3, 3, 4, 5. What is the probability of rolling an odd number?

4. A bag contains 6 red balls, 8 green balls and 10 blue balls. What is the probability that a ball chosen at random is neither red nor blue?

5. The school carpark contains 64 cars. Four cars were driven by Year 11 students and five times as many cars were driven by Year 12 students as Year 11 students. The remainder of the cars were driven by staff at the school. If a car is chosen at random, what is the probability that it is driven by a staff member?

6. CHALLENGE In a tennis club there are 8 more females than males. If a person is selected from the members of the club, the probability that the person is male is $\frac{3}{7}$. How many female members are in the club?

Answers page 161

KEY SKILL

49 Probability B

HINTS

- Make sure you know the **Probability Hints** from Key Skill 48 on page 118.
- Make sure you know the **Equations Hints** from Key Skill 26 on page 68.

Reminder!

- Read the question carefully to identify what needs to be found.
- Re-read the question when you've finished your answer to make sure that the question has been answered and your solution makes sense.

Examples

Three friends play a round of golf. Jack is twice as likely to win as Brad and Lee. If Brad and Lee have an equal chance of winning, find the probability that Lee wins the round.

Solution

Let P(Brad wins) $= x$

$\therefore$ P(Lee wins) $= x$ and P(Jack wins) $= 2x$

$$\therefore x + x + 2x = 1$$
$$4x = 1$$
$$x = \frac{1}{4}$$

$\therefore$ the probability that Lee wins is $\frac{1}{4}$.

FOCUS on ...

1. The question: Asks you to find the probability Lee wins the round.
2. The information: Gives you the likelihood of three friends winning the round.
3. Your working: You need to introduce a pronumeral to set up an equation to solve.
   Let P(Brad wins) = x.
   Express the probability of Lee and Jack in terms of x:
   P(Lee wins) = x and P(Jack wins) = 2x.
   The sum of the three probabilities equals 1:
   $$x + x + 2x = 1$$
   $$4x = 1$$
   $$x = 0.25$$
4. Your answer: The probability that Lee wins is 0.25.

A box contains red, blue and green balls. The chance of selecting a green ball is twice the chance of selecting a blue ball which is twice the chance of selecting a red ball. What is the probability of selecting a blue ball?

Solution

Let P(red) $= x$

$\therefore$ P(blue) $= 2x$ and P(green) $= 4x$

$$x + 2x + 4x = 1$$
$$7x = 1$$
$$x = \frac{1}{7}$$

$\therefore$ as P(blue) $= 2x$, then the probability of choosing a blue ball is $\frac{2}{7}$.

FOCUS on ...

1. The question: Asks you to find the probability of selecting a blue ball.
2. The information: Gives you the likelihood of three different colour balls being selected from a box.
3. Your working: You need to introduce a pronumeral to set up an equation to solve. Let P(red) = x.
   Express the probability of blue and green balls being selected in terms of x.
   P(blue) = 2x and P(green) = 4x
   The sum of the three probabilities equals 1.
   $$x + 2x + 4x = 1$$
   $$7x = 1$$
   $$x = \frac{1}{7}$$
   Multiply by 2 to find the value of 2x: P(blue) = $\frac{2}{7}$.
4. Your answer: The probability of blue is $\frac{2}{7}$.

Now try these!

1. Kayla, Julian and Dianne are to play a round of golf. Kayla is twice as likely to win as Julian and Julian is three times more likely to win than Dianne. What is the probability that Dianne will win?

2. A bag contains twice as many red balls as blue balls and there are four times as many blue balls as pink balls. If a ball is chosen at random find the probability it is a blue ball.

3. A bag of Chockys has different coloured buttons of chocolate. There are red, blue, green and yellow buttons. The chance of selecting a red is twice as likely as choosing a blue, green or yellow. What is the probability of selecting a red?

4. George enters a chess tournament and his chance of winning is three times as likely as any other player. If there is a total of 8 competitors, what is the probability that George wins?

5. Four tennis players remain in a knockout tournament. Joey is twice as likely to win the final as Matt. Sam is half as likely to win as Matt and twice as likely to win as David. What is the probability that Matt will win the tournament?

6. CHALLENGE A bag contains red frogs, green frogs and black frogs. There are four times as many red frogs as black frogs and half as many black frogs as green frogs. Anna selected a frog at random from the bag, noted that it was red and ate it. If 34 frogs remain in the bag, what is the chance that the next frog chosen at random will be red?

Answers pages 161–162

KEY SKILL

50 Probability C

HINTS

- Data can be displayed in a Venn Diagram or a Two-way Table and can be used to answer probability questions,
  e.g. Students were surveyed as to their enjoyment of Science and History and the results recorded in the diagrams below:

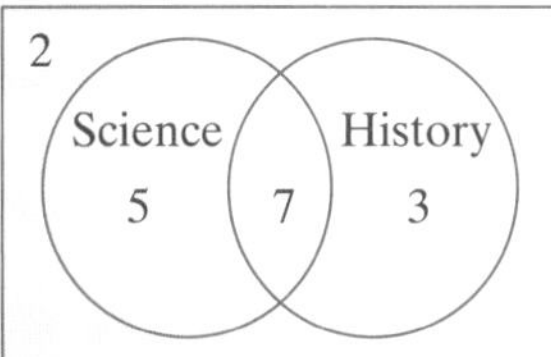

	Enjoy History	Do not enjoy History
Enjoy Science	7	5
Do not enjoy Science	3	2

If a student is chosen at random, what is the probability that they like both subjects? $\frac{7}{17}$

Reminder!

- Read the question carefully to identify what needs to be found.
- Re-read the question when you've finished your answer to make sure that the question has been answered and your solution makes sense.

Examples

A group of 30 students were surveyed. There were 12 students choosing Food Technology and 15 students choosing Media. Eight students were not choosing either subject. Use a Venn diagram to find the probability that a student selected at random had chosen both subjects?

Solution

Use F: Food Technology and M: Media

P(both) $= \frac{5}{30} = \frac{1}{6}$

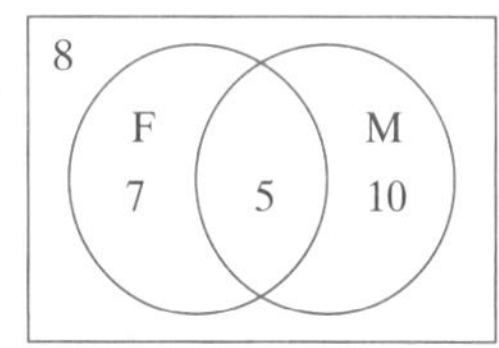

FOCUS on ...

1. **The question:** Asks you to find the probability that a student had chosen both subjects.
2. **The information:** Gives you the number who had chosen Food Technology, chosen Media and chosen neither.
3. **Your working:** Draw a Venn Diagram and write 8 outside the circles.
   Subtract those who chose neither from total students.
   $30 - 8 = 22$ means 22 students chose at least one of the subjects.
   Add those who had chosen each subject and subtract the number who had chosen at least one.
   $12 + 15 - 22 = 5$ means 5 out of 30 students chose both.
   Write the probability as $\frac{5}{30} = \frac{1}{6}$.
4. **Your answer:** The probability is $\frac{1}{6}$.

In the elective music class, 15 students are in the band, 18 students are in the choir, 12 are in both and 4 are in neither. Use a two-way table to find the probability that a student selected at random is in the choir but not in the band.

Solution

	In band	Not in band
In choir	12	6
Not in choir	3	4

P(in choir but not band) $= \frac{6}{25}$

FOCUS on ...

1. **The question:** Asks you to find the probability that a student is in the choir but not in the band.
2. **The information:** Gives you the number in the band, in the choir, in both and neither.
3. **Your working:** Draw a two-way table and write 4 in the not band/not choir entry.
   Write 12 in the both band and choir entry.
   Subtract those who are in both from those who are in the choir: $18 - 12 = 6$ means 6 are in the choir but not the band.
   Subtract those who are in both from those who are in the band: $13 - 12 = 3$ means 3 are in the band but not the choir.
   Total all entries: $12 + 3 + 6 + 4 = 25$ means 25 students.
   Write the probability as $\frac{6}{25}$.
4. **Your answer:** The probability is $\frac{6}{25}$.

Now try these!

1. In a survey of 100 male students it was found that 21 students play football and not cricket, 19 students play cricket and not football, and 27 students play neither. Use a Venn diagram to find the probability that a boy chosen at random plays both sports.

2. The manager at a restaurant conducted a customer satisfaction survey. At lunchtime, 28 customers were pleased with the service and 10 were disappointed. There were 43 dinner customers and 18 reported that they were disappointed with the service. Use a two-way table to find the probability that a customer chosen at random was a dinner guest who was pleased with the restaurant's service.

3. In a survey of 350 teenage customers at a large shopping centre, 250 said they shop at Jaycees, 220 shop at UShop and 190 shop at both stores. Use a Venn diagram to find the percentage of customers who do not shop in either store.

4. When semester reports were distributed in a Year 8 class, it was found that 10 students received at least one grade A, 14 students received at least one B, and 6 students received As and Bs. One-quarter of the students in the class received no As or Bs. Use a two-way table to find the number of students who did not receive an A or a B.

5. A survey of 48 students whose parents are of Chinese background found that one-quarter of the students speak both Cantonese and Mandarin, half the students speak Mandarin and one-twelfth of the students speak neither language. If a student is chosen at random, what is the probability that they only speak Cantonese?

6. CHALLENGE The results of a survey of 60 people showed that the ratio of the number of people who drank both tea and coffee to the number of people who drank neither was 2 : 1. 20% of the people only drank tea and 30% only drank coffee. Complete a Venn diagram to find the probability that a person chosen at random drank both tea and coffee.

Answers pages 162–163

REVISION TEST 15 Level of difficulty—Average

1. The ages of 4 children in a family are recorded. The median age of the children is 6 and the range of ages is 9. The ages arranged in order are 3, 4, x, y. What are the ages of the two oldest children?

2. A bag contains 4 green gumdrops and 3 yellow gumdrops. Karl pulls a yellow gumdrop out of the bag and eats it. What is the probability that the next gumdrop chosen at random will also be yellow?

3. Abby writes the numbers 1 to 20 on identical cards and places them in a bag. A card is chosen at random. What is the probability that Abby chooses a card that is less than 16, odd and composite?

4. Lee rolls a six-sided die once. What is the probability that the number he rolls is an even number or less than 3?

5. The mean mass of 3 dogs is 14 kg. One of the dogs, Max, has a mass of 18 kg. The other two dogs, Bella and Jack, have the same mass. What is Bella's mass?

6. At Paul's Fish and Chips shop 60 customers who ordered chips were surveyed. Thirty customers wanted salt and vinegar, 16 wanted salt only and 12 did not want salt or vinegar. If a customer is chosen at random, what is the probability that they wanted vinegar but no salt?

Answers page 163

REVISION TEST 16 Level of difficulty—Challenging

1. George recorded the results of 4 science quizzes, marked out of 20. His results were 12, 16, 14 and 10. What does he need to score in the next test to increase his mean by one?

2. The probability that each of the players in the semi-finals of a tournament will win the final has been calculated. Sam is twice as likely to win as Oscar. Winston is as likely to win as Sam but three times more likely to win than Cody. What is the probability that Oscar will win the final?

3. The composite numbers between 20 and 30 are written on balls and placed into a bag. One ball is chosen at random. If the selected ball is an odd number, what is the probability that it is a multiple of 3?

4. The mode of 5 scores is half the median. The mean of the scores is half the range. The scores arranged in ascending order are 4, 4, x, y, 24. What are the two missing numbers?

5. A survey of 20 people was used to find the number of people who had travelled overseas. There were 12 females involved in the survey. Five males and 8 females had travelled overseas. If a person was chosen at random, what is the probability that they were a male who had not travelled overseas?

6. In a certain school, 30% of Year 12 is studying Physics, 25% are studying Chemistry and 15% are studying both. There are 72 students who are studying neither Physics nor Chemistry. How many students in Year 12 are studying exactly one of these science subjects?

Answers page 164

Worked solutions

NUMBER

Key Skill  Integers (pages 10–11)

1. **22°**

$$\begin{aligned}\text{Range} &= 9 - -13\\ &= 9 + 13\\ &= 22\end{aligned}$$

∴ the temperature range was 22°.

2. **43**

$$\begin{aligned}\text{Result} &= 14 \times 5 + 8 \times (-3) + 3 \times (-1)\\ &= 70 - 24 - 3\\ &= 43\end{aligned}$$

∴ Yun scored 43.

3. **208°**

$$\begin{aligned}\text{Difference} &= -2 - -210\\ &= -2 + 210\\ &= 208\end{aligned}$$

∴ the difference is 208°.

4. **−17°**

$$\begin{aligned}\text{Temperature} &= -1 + 10 \times (-1) + 3 \times (-2)\\ &= -1 - 10 - 6\\ &= -17\end{aligned}$$

∴ the temperature is −17°.

5. **−10**

$$\begin{aligned}\text{Result} &= 2 \times 3 + 3 \times (-3) + 1 \times (-5) + 2 \times (-1)\\ &= 6 - 9 - 5 - 2\\ &= -10\end{aligned}$$

∴ Amanda scored −10.

6. **−40 °F**

$$\begin{aligned}F &= \frac{9}{5}C + 32\\ &= \frac{9}{5} \times (-40) + 32\\ &= -72 + 32\\ &= -40\end{aligned}$$

∴ the temperature is −40 °F.

Key Skill  Unitary method A (pages 12–13)

1. **6 h**

$$\begin{aligned}\text{Time for 1 bricklayer} &= 8 \times 3\\ &= 24\\ \text{Time for 4 bricklayers} &= 24 \div 4\\ &= 6\end{aligned}$$

∴ it would take 6 h.

2. **3 days**

$$\begin{aligned}\text{Time for 1 dog} &= 4 \times 3\\ &= 12\\ \text{Time for 4 dogs} &= 12 \div 4\\ &= 3\end{aligned}$$

∴ a bag would last 3 days.

3. **18 days**

$$\begin{aligned}\text{Time for 1 woman} &= 6 \times 6\\ &= 36\\ \text{Time for 2 women} &= 36 \div 2\\ &= 18\end{aligned}$$

∴ the job would take 18 days.

4. **64 machines**

$$\begin{aligned}\text{Machines for 1 day} &= 8 \times 4\\ &= 32\\ \text{Machines for } \tfrac{1}{2} \text{ day} &= 32 \div \frac{1}{2}\\ &= 64\end{aligned}$$

∴ 64 machines are needed.

5. **18 days**

$$\begin{aligned}\text{Total food-days for 1 soldier} &= 72 \times 20\\ &= 1440\\ \text{Total food-days for 80 soldiers} &= 1440 \div 80\\ &= 18\end{aligned}$$

∴ the food will last another 18 days.

6. $\mathbf{10\frac{1}{2}}$ **days**

$$\begin{aligned}\text{Days for 1 man} &= 6 \times 8\\ &= 48\\ \text{Days remaining} &= 48 - 3 \times 6\\ &= 30\\ \text{Days remaining for 4 men} &= 30 \div 4\\ &= 7\frac{1}{2}\\ \text{Total time} &= 7\frac{1}{2} + 3\\ &= 10\frac{1}{2}\end{aligned}$$

∴ it took a total of $10\frac{1}{2}$ days for the job.

Key Skill  Unitary method B (pages 14–15)

1. **2 h**

LCM of 3 and 6: 6

Kirra: fences in 6 h = 2

Bradie: fences in 6 h = 1

Kirra and Bradie: fences in 6 h = 3

As 6 ÷ 3 = 2, then it will take 2 h.

2. **18 min 45 s**

LCM of 30 and 50: 150

Pipe A: tanks in 150 min = 5

Pipe B: tanks in 150 min = 3

Pipes A and B: tanks in 150 min = 8
As 150 ÷ 8 = 18.75, then it will take 18.75 min, or 18 min 45 s.

3. **4 min 48 s**
LCM of 8 and 12: 24
Jason: classrooms in 24 min = 3
Seth: classrooms in 24 min = 2
Jason and Seth: classrooms in 24 min = 5
As 24 ÷ 5 = 4.8, then it will take 4.8 min, or 4 min 48 s.

4. **6 min 40 s**
LCM of 15 and 12: 60
Employee 1: shelves in 60 min = 4
Employee 2: shelves in 60 min = 5
Both employees: shelves in 60 min = 9
As $60 \div 9 = 6\frac{2}{3}$, then it will take $6\frac{2}{3}$ min, or 6 min 40 s.

5. **$1\frac{5}{7}$ h**
LCM of 4 and 3: 12
Lee: stables in 12 h = 3
Beck: stables in 12 h = 4
Lee and Beck: stables in 12 h = 7
As $12 \div 7 = 1\frac{5}{7}$, then it will take $1\frac{5}{7}$ h.

6. **5 pm**
LCM of 4 and 2: 4
Hannah: rooms in 4 h = 1
Mum: rooms in 4 h = 2
Hannah and Mum: rooms in 4 h = 3
As $4 \div 3 = 1\frac{1}{3}$, then it will take $1\frac{1}{3}$ h if they work together.
But only $\frac{3}{4}$ of a room to clean: $\frac{3}{4} \times 1\frac{1}{3} = 1$
It will take another hour.
∴ they will finish at 5 pm.

Key Skill  Unitary method C

(pages 16–17)

1. **4 h**
LCM of 80 and 120: 240
Both men: jobs in 240 min = 3
First man: jobs in 240 min = 2
As 3 − 2 = 1, then the second man takes 240 min to complete one job.
∴ it would take the second man 240 min, or 4 h.

2. **9 min**
It takes 18 min for hot water tap.
LCM of 6 and 18: 18
Both taps: baths in 18 min = 3
Hot water tap: baths in 18 min = 1
As 3 − 1 = 2, then the cold water fills 2 baths in 18 min, or 1 bath in 9 min.
∴ it would take 9 min.

3. **1 h 12 min**
One and a half hours = 90 min
LCM of 90 and 40: 360
Megan: lawns in 360 min = 4
Both: lawns in 360 min = 9
As 9 − 4 = 5, then Hayley rakes 5 lawns in 360 min, or 1 lawn in 72 min, or 1 h 12 min.
∴ it would take Hayley 1 h 12 min.

4. **20 min**
LCM of 30 and 12: 60
Miles: boxes in 60 min = 2
Both: boxes in 60 min = 5
Millie can deliver 3 boxes in 60 min, or one box in 20 min.
∴ it would take Millie 20 min.

5. **7 h 30 min**
LCM of 3 and 5: 15
Both: jobs in 15 h = 5
Andrew: jobs in 15 h = 3
Rose can complete 2 jobs in 15 h, or one job in 7.5 h.
∴ it would take Rose 7 h 30 min.

6. **30 min**
Half job in 6 min is whole job in 12 min.
LCM of 20 and 12: 60
Donald: jobs in 60 min = 3
Both: jobs in 60 min = 5
Daffy can complete 2 jobs in 60 min, or one job in 30 min.
∴ it would take Daffy 30 min.

Key Skill 5 Fractions with unitary method A

(pages 18–19)

1. **2 km**

$$\begin{aligned}\text{Two-fifths of distance} &= 1.2\\ \text{Five-fifths of distance} &= 1.2 \div 2 \times 5\\ &= 3\\ \text{Two-thirds of distance} &= \frac{2}{3} \times 3\\ &= 2\end{aligned}$$

∴ James walked 2 km.

2. **70**

$$1 - \frac{3}{8} = \frac{5}{8}$$

$$\begin{aligned}\text{Five-eighths of book} &= 175\\ \text{Eight-eighths of book} &= 175 \div 5 \times 8\\ &= 280\\ \text{One-quarter of book} &= \frac{1}{4} \times 280\\ &= 70\end{aligned}$$

∴ Bridget will have 70 pages remaining.

3. **$1.90**

$$\begin{aligned}\text{Two-fifths of allowance} &= 7.6\\ \text{Five-fifths of allowance} &= 7.6 \div 2 \times 5\\ &= 19\\ \text{One-tenth of allowance} &= \frac{1}{10} \times 19\\ &= 1.9\end{aligned}$$

$\therefore$ Phelia saves $1.90 each week.

4. **4480 L**

$$\begin{aligned}\text{Three-twentieths full} &= 840\\ \text{Twenty-twentieths full} &= 840 \div 3 \times 20\\ &= 5600\\ \text{Four-fifths of full} &= \frac{4}{5} \times 5600\\ &= 4480\end{aligned}$$

$\therefore$ the tank contains 4480 L.

5. **2 km**

$$\begin{aligned}\text{Distance run} &= 1 - (\frac{2}{5} + \frac{1}{4})\\ &= \frac{7}{20}\\ \text{Seven-twentieths of dist.} &= 2.8\\ \text{Twenty-twentieths of dist.} &= 2.8 \div 7 \times 20\\ &= 8\\ \text{One-quarter of dist.} &= \frac{1}{4} \times 8\\ &= 2\end{aligned}$$

$\therefore$ Mustaf walks 2 km.

6. **720**

$$\begin{aligned}\text{Harley PS sport players} &= \frac{2}{3} \times 240\\ &= 160\\ \text{Two-ninths of Harley HS} &= 160\\ \text{Nine-ninths of Harley HS} &= 160 \div 2 \times 9\\ &= 720\end{aligned}$$

$\therefore$ 720 students attend Harley HS.

Key Skill  6 Fractions with unitary method B

(pages 20–21)

1. **$160**

$$\begin{aligned}\text{Fraction for earrings} &= \frac{3}{4} \times (1 - \frac{2}{3})\\ &= \frac{1}{4}\\ \text{One-quarter of birthday money} &= 40\\ \text{Four-quarters of birthday money} &= 40 \times 4\\ &= 160\end{aligned}$$

$\therefore$ Jessica received $160 for her birthday.

2. **960 km**

$$\begin{aligned}\text{Fraction remaining after day 1} &= \frac{2}{3}\\ \text{Fraction on day 2} &= \frac{3}{4} \times \frac{2}{3}\\ &= \frac{1}{2}\\ \text{Half total distance} &= 480\\ \text{Whole distance} &= 960\end{aligned}$$

$\therefore$ total distance of trip is 960 km.

3. **$1440**

$$\begin{aligned}\text{Fraction after World Vision} &= \frac{1}{4}\\ \text{Fraction to Red Cross} &= \frac{1}{3} \times \frac{1}{4}\\ &= \frac{1}{12}\\ \text{One-twelfth of total money} &= 120\\ \text{Twelve-twelfths of money} &= 120 \times 12\\ &= 1440\end{aligned}$$

$\therefore$ Seth received $1440 inheritance.

4. **24**

$$\begin{aligned}\text{Fraction that are red} &= \frac{1}{4}\\ \text{Fraction not red} &= \frac{3}{4}\\ \text{Fraction that are black} &= \frac{1}{3} \times \frac{3}{4}\\ &= \frac{1}{4}\\ \text{One-quarter of all jelly beans} &= 6\\ \text{Four-quarters of all jelly beans} &= 24\end{aligned}$$

$\therefore$ there are 24 jelly beans in the bag.

5. **90**

$$\begin{aligned}\text{Fraction that are schoolfriends} &= \frac{2}{3}\\ \text{Fraction not schoolfriends} &= \frac{1}{3}\\ \text{Fraction that are relatives} &= \frac{3}{5} \times \frac{1}{3}\\ &= \frac{1}{5}\\ \text{One-fifth of all friends} &= 18\\ \text{Five-fifths of all friends} &= 90\end{aligned}$$

$\therefore$ Brittany has 90 friends on the site.

6. **$640**

$$\begin{aligned}\text{Fraction to parents and sister} &= \frac{1}{4} + \frac{1}{5}\\ &= \frac{9}{20}\\ \text{Fraction not to parents or sister} &= \frac{11}{20}\\ \text{Fraction to brother} &= \frac{1}{11} \times \frac{11}{20}\\ &= \frac{1}{20}\\ \text{One-twentieth of money} &= 32\\ \text{Twenty-twentieths of money} &= 640\end{aligned}$$

$\therefore$ Julia found $640.

Key Skill 7 Fractions with unitary method C

(pages 22–23)

1. **240 mL**

$\text{Fraction used} = \frac{2}{3} - \frac{1}{2} = \frac{1}{6}$

One-sixth of total soap = 40
Six-sixths of total soap = 240
∴ dispenser holds 240 mL of soap.

2. **5 km**

$\text{Difference in fractions} = \frac{2}{3} - \frac{1}{4} = \frac{5}{12}$

Five-twelfths of distance = 25
Twelve-twelfths of distance = 25 ÷ 5 × 12 = 60

$\text{Fraction swimming} = 1 - (\frac{2}{3} + \frac{1}{4}) = \frac{1}{12}$

$\text{Swim distance} = \frac{1}{12} \times 60 = 5$

∴ Sigrid swims 5 km.

3. **1800 L**

$\text{Fraction filled} = \frac{4}{5} - \frac{1}{2} = \frac{3}{10}$

Three-tenths of pool = 540
Ten-tenths of pool = 1800
∴ the pool can hold 1800 L.

4. **$3150**

$\text{Fraction spent in that week} = \frac{2}{3} - \frac{3}{10} = \frac{11}{30}$

Eleven-thirtieths of savings = 1155
Thirty-thirtieths of savings = 1155 ÷ 11 × 30 = 3150
∴ Donika had saved $3150.

5. **800 km**

$\text{Middle fraction} = 1 - (\frac{1}{4} + \frac{2}{5}) = \frac{7}{20}$

Seven-twentieths of distance = 280
Total distance = 280 ÷ 7 × 20 = 800
∴ the distance is 800 km.

6. **$3600**

$\text{Year 8 fraction} = 1 - (\frac{1}{6} + \frac{3}{8}) = \frac{22}{48} = \frac{11}{24}$

Eleven-twenty-fourths of total = 4400
Total raised = 4400 ÷ 11 × 24 = 9600

$\text{Year 9 amount} = \frac{3}{8} \times 9600 = 3600$

∴ Year 9 raised $3600.

Revision Test  Level of difficulty—Average

(page 24)

1. **−5**

Score = 3 × 4 + 1 × (−3) + 2 × (−4) + 3 × (−2)
= 12 − 3 − 8 − 6
= −5
∴ Natasha scored −5.

2. **2232 kg**

Amount of food for 6 days = 432
Amount of food for 31 days = 432 ÷ 6 × 31 = 2232
∴ elephant will eat 2232 kg.

3. **16 min**

Time for 3 hoses = 40
Time for 1 hose = 40 × 3 = 120
Time for 5 hoses = 120 ÷ 5 = 24
Now, 40 − 24 = 16.
∴ 16 min would be saved.

4. **100**

One-quarter of total = 60
Four-quarters of total = 240

$\text{Choc-chip} = \frac{1}{3} \times 240 = 80$

Lemon-poppyseed = 240 − (60 + 80) = 100
∴ 100 lemon-poppyseed muffins were made.

5. **7.2 m**

$\text{Fraction painted green} = 1 - (\frac{1}{2} + \frac{1}{3}) = \frac{1}{6}$

One-sixth of the pole = 1.2
Six-sixths of pole = 1.2 × 6 = 7.2
∴ the pole is 7.2 m long.

6. €1800

Fraction spent in middle week $= 1 - (\frac{2}{5} + \frac{1}{4})$
$= \frac{7}{20}$

Seven-twentieths of money = 630
Twenty-twentieths of money = 630 ÷ 7 × 20
= 1800

∴ Fiona originally had €1800.

Revision Test 2
Level of difficulty—Challenging (page 25)

1. 10

Days for 5 workers = 6
Days for 1 worker = 30
∴ the job will take 30 work-days.
After 2 days, 10 work-days have been used.
∴ there are still 20 work-days required.
As 20 ÷ 2 = 10, then it will take the 2 workers another 10 days.
∴ it will take 10 more days.

2. 5:18

Mickayla: Rooms in 2 h = 1
Mother: Rooms in 2 h = 4
Both: Rooms in 2 h = 5
As 120 ÷ 5 = 24, when they work together, a room is cleaned in 24 min.
By 5:00 pm, the room is 'one-quarter clean'.
As $\frac{3}{4} \times 24 = 18$, then another 18 min is needed.
∴ the room is clean at 5:18.

3. one-third

LCM of 4 and 6: 12
Both: Number of fences in 12 h = 3
Jack: Number of fences in 12 h = 2
∴ Jill can paint a fence in 12 h.
As Jack paints a fence in half the time as Jill, then together Jill completes a third of the job and Jack two-thirds.
∴ Jill does one-third of the job.

4. 50

Michael: Number of finches $= 45 \times \frac{2}{3}$
= 30
Scott: Three-fifths of birds = 30
Five-fifths of birds = 30 ÷ 3 × 5
= 50

∴ Scott has 50 birds.

5. 90

Left handed girls $= 1 - \frac{5}{6}$
$= \frac{1}{6}$
Brown-haired left-handed girls $= \frac{2}{3} \times \frac{1}{6}$
$= \frac{1}{9}$
Brown-haired left-handed female tennis players
$= \frac{1}{5} \times \frac{1}{9}$
$= \frac{1}{45}$
One-forty-fifth of girls = 2
Total number of girls = 2 × 45
= 90
∴ there are 90 girls in Year 8.

6. 6

Two-fifths of book = 288
Five-fifths of book = 288 ÷ 2 × 5
= 720
Fraction remaining $= 1 - (\frac{3}{10} + \frac{1}{8} + \frac{2}{5} + \frac{1}{6})$
$= \frac{1}{120}$
Pages for Friday $= \frac{1}{120} \times 720$
= 6
∴ Isabella has 6 pages to read Friday.

Key Skill 8 Percentages and unitary method
(pages 26–27)

1. \$250

24% of total bill = 60
100% of total bill = 60 ÷ 24 × 100
= 250
∴ the total cost was \$250.

2. 54

20% of students = 45
100% of students = 45 × 5
= 225
24% of students = 0.24 × 225
= 54
∴ 54 students chose Outdoor Recreation.

3. 80

85% of total marks = 68
100% of total marks = 68 ÷ 85 × 100
= 80
∴ the test was out of 80.

4. 67.5 kg

8% of total mass = 6
100% of total mass = 6 ÷ 8 × 100
= 75
90% of total mass = 0.9 × 75
= 67.5
∴ Liam's mass is 67.5 kg.

5. 7 kg

40% of total mass = 8
100% of total mass = 8 ÷ 40 × 100
= 20
% copper = 100 − (25 + 40)
= 35
Mass of copper = 0.35 × 20
= 7
∴ there is 7 kg of copper.

6. 184.5 cents/L

25 − 22.5 = 2.5
2.5% of old price = 4.5
100% of old price = 4.5 ÷ 2.5 × 100
= 180
New price = 180 + 4.5
= 184.5
∴ petrol will cost 184.5 cents/L.

Key Skill  Increasing and decreasing percentages A

(pages 28–29)

1. 912 g

Mass of new box = 1.2 × 760
= 912
∴ the new mass is 912 g.

2. $90

75% of original price = 67.5
100% of original price = 67.5 ÷ 75 × 100
= 90
∴ the original price was $90.

3. 12 760

New crowd = 0.88 × 14 500
= 12 760
∴ the new crowd was 12 760.

4. $720 000

112% of Feb. sales = 806 400
100% of Feb. sales = 806 400 ÷ 112 × 100
= 720 000
∴ February sales totalled $720 000.

5. $250

80% of original price = 200
100% of original price = 200 ÷ 80 × 100
= 250
∴ the bicycle originally cost $250.

6. $2.40

Eric: 75% of RRP = 18
100% of RRP = 18 ÷ 75 × 100
= 24
Also, 35 − 25 = 10
∴ 10% difference in price
10% of RRP = 0.1 × 24
= 2.4
∴ there is a difference of $2.40.

Key Skill  Increasing and decreasing percentages B

(pages 30–31)

1. $4.80

New price = 48 × 1.2 × 0.75
= 43.2
Difference = 48 − 43.2
= 4.8
∴ there was a price drop of $4.80.

2. 4.94%

0.97 × 0.98 = 0.9506
Decrease = 1 − 0.9506
= 0.0494
∴ his mass dropped by 4.94%.

3. $15

New allowance = 10 × 1.2 × 1.25
= 15
∴ her allowance is $15.

4. 1.2%

0.92 × 1.1 = 1.012
Increase = 1.012 − 1
= 0.012
∴ there was a population increase of 1.2%.

5. $69.85

New price = 59 × 1.1 × 1.05 × 1.025
= 69.848 625
= 69.85 (2 dec. pl.)
∴ the price is $69.85.

6. Choice 2

Choice 1: 500 × 1.03 × 3
= 1545
Choice 2: 500 × 1.01 + 500 × 1.01 × 1.02
+ 500 × 1.01 × 1.02 × 1.03
= 1550.653
= 1550.65 (2 dec. pl.)
∴ the second choice gives $5.65 more.

Key Skill  Financial mathematics: Profit and loss A (pages 32–33)

1. **55%**

$$\text{Profit} = 186\,000 - 120\,000 = 66\,000$$
$$\text{Profit \%} = \frac{66\,000}{120\,000} \times 100\% = 55$$

∴ the profit is 55%.

2. **$440**

$$\text{Loss} = 45\% \text{ of } 800 = 0.45 \times 800 = 360$$

∴ the loss was $360.

$$\text{Selling price} = \text{Cost price} - \text{Loss} = 800 - 360 = 440$$

∴ the dog was sold for $440.

3. **$33.\dot{3}\%$**

$$\text{Cost} = 660 + 240 = 900$$
$$\text{Profit} = 1200 - 900 = 300$$
$$\text{Profit \%} = \frac{300}{900} \times 100\% = 33.333\,3333\ldots = 33.\dot{3}$$

∴ the profit is $33.\dot{3}\%$.

4. **56.25%**

Consider the LCM of 4 and 5 = 20

Cost = 5 apples for $4 is 20 apples for $16

Sell = 4 apples for $5 is 20 apples for $25

$$\text{Profit on 20 apples} = 25 - 16 = 9$$
$$\text{Profit \%} = \frac{9}{16} \times 100\% = 56.25$$

∴ the profit is 56.25%.

5. **$150**

Cost : Sell = 4 : 5

As 5 − 4 = 1, then profit % $= \frac{1}{4} \times 100\% = 25\%$

$$\text{Selling price} = 1.25 \times 120 = 150$$

∴ the tripod sold for $150.

6. **$7.70/kg**

$$\text{Total cost} = 30 \times 4 + 20 \times 5 = 220$$
$$\text{Total mass} = 30 + 20 = 50$$
$$\text{Cost/kg} = 220 \div 50 = 4.4$$
$$\text{Sales price} = 4.4 \times 1.75 = 7.7$$

∴ the mixed nuts will sell for $7.70/kg.

Key Skill  Financial mathematics: Profit and loss B (pages 34–35)

1. **$89.60**

125% of cost price = 80

140% of cost price = 80 ÷ 125 × 140 = 89.6

∴ the pair of shoes should be sold for $89.60.

2. **$36**

140% of cost price = 56

90% of cost price = 56 ÷ 140 × 90 = 36

∴ the tree is now priced at $36.

3. **$21**

120% of cost price = 24

105% of cost price = 24 ÷ 120 × 105 = 21

∴ the toaster will sell for $21.

4. **$50**

180% of cost price = 75

New price = 75 ÷ 180 × 120 = 50

∴ the bag will be priced at $50.

5. **$41.25**

160% of cost price = 80

Easter price = 80 ÷ 160 × 110 = 55

Discounted price = 0.75 × 55 = 41.25

∴ the calligraphy set will sell for $41.25.

6. **$30 000**

Profit % = 0.8 × 1.4 = 1.12

∴ dealership makes a 12% profit

12% of cost price = 3600

100% of cost price = 3600 ÷ 12 × 100 = 30 000

∴ the car cost the dealership $30 000.

Key Skill  Financial mathematics—GST (pages 36–37)

1. **$264**

Charge = 80 × 1.1 × 3 = $264

∴ the cost is $264.

2. **$5.75**
New price = 5.5 ÷ 110 × 115
= 5.75
∴ the paper costs $5.75.

3. **$38**
GST on labour = 65 × 0.1 × 4
= 26
GST on products = 132 ÷ 11
= 12
Total GST = 26 + 12
= 38
∴ the total GST was $38.

4. **$12.73**
Original price = 840 ÷ 120 × 100
= 700
Price increase = 840 − 700
= 140
Amount of GST = 140 ÷ 11
= 12.727 272
= 12.73 (2 dec. pl.)
∴ the GST increased by $12.73.

5. **$72.60**
2% of original price without GST = 1.32
Original price without GST = 1.32 ÷ 2 × 100
= 66
Original price with GST = 66 × 1.1
= 72.6
∴ the original price is $72.60.

6. **$3097.60**
GST drop = 20% of (original price ÷ 11) = 70.4
$\frac{0.2}{11}$ of original price = 70.4
0.2 of original price = 70.4 × 11
= 774.4
Original price = 774.4 ÷ 0.2
= 3872
Discounted price = 0.8 × 3872
= 3097.6
∴ the discounted price is $3097.60.

Revision Test 3
Level of difficulty—Average (page 38)

1. **24 min**
Percentage in pool = 100 − (40 + 35)
= 25
∴ 25% of time swimming
25% of time = 15
40% of time = 15 ÷ 25 × 40
= 24
∴ Sarah was on the treadmill for 24 min.

2. **$120**
85% of original price = 102
100% of original price = 102 ÷ 85 × 100
= 120
∴ original price was $120.

3. **$226 812.50**
Value in 2015 = 190 000 × 0.955 × 1.25
= 226 812.50
∴ the block is valued at $226 812.50.

4. **$16 250**
New price = 0.65 × 25 000
= 16 250
∴ Tom sold the bike for $16 250.

5. **$30**
140% of cost price = 840
Cost price = 840 ÷ 140 × 100
= 600
∴ television cost price is $600
Discounted price = 0.75 × 840
= 630
630 − 600 = 30
∴ the store will make $30 profit.

6. **$0.22**
GST on $12 = 12 ÷ 11
= 1.090 909 09 …
= 1.09 (2 dec. pl.)
Discounted price = 0.8 × 12
= 9.6
GST on $9.60 = 9.6 ÷ 11
= 0.872 727 2727 …
= 0.87 (2 dec. pl.)
1.09 − 0.87 = 0.22
∴ GST drops by $0.22.
[Alternatively, 20% of $1.09 = $0.22.]

Revision Test 4
Level of difficulty—Challenging (page 39)

1. **$1225**
2.5% off original price = 35
100% of original price = 35 ÷ 2.5 × 100
= 1400
New price = 0.875 × 1400
= 1225
∴ the new price is $1225.

2. **$21**
Laddies: Price = 0.85 × 420
= 357
Difference = 357 − 336
= 21
∴ Ed's price is $21 cheaper than Laddies'.

3. **40%**
0.75 × 0.8 = 0.6
∴ consecutive discounts is the same as a 40% discount.

4. **\$8.24/kg**

$$\begin{aligned}\text{Total cost} &= 5.6 \times 25 + 4.9 \times 45\\ &= 360.5\\ \text{Total mass} &= 25 + 45\\ &= 70\\ \text{Cost/kg} &= 360.5 \div 70\\ &= 5.15\\ \text{Sale price} &= 5.15 \times 1.6\\ &= 8.24\end{aligned}$$

∴ the mixture should sell for \$8.24/kg.

5. **13.79%**

$$\begin{aligned}116\% \text{ of cost price} &= 1392\\ 16\% \text{ of cost price} &= 1392 \div 116 \times 16\\ &= 192\end{aligned}$$

∴ the profit was \$192.

$$\begin{aligned}\text{Profit \% of SP} &= \frac{192}{1392} \times 100\%\\ &= 13.793\,1034\ldots\\ &= 13.79 \text{ (2 dec. pl.)}\end{aligned}$$

∴ the profit was 13.79% of the sale price.

6. **\$1100**

$$\begin{aligned}\text{New GST} &= 85\\ \text{New price} &= 85 \times 11\\ &= 935\\ 85\% \text{ of old price} &= 935\\ 100\% \text{ of old price} &= 935 \div 85 \times 100\\ &= 1100\end{aligned}$$

∴ the original price was \$1100.

Key Skill  14 Ratio and rates: Simplifying ratios A

(pages 40–41)

1. **19 : 5**

$$\begin{aligned}\text{No. unfinancial} &= 1200 - 950\\ &= 250\\ \text{Ratio} &= 950 : 250\\ &= 19 : 5\end{aligned}$$

∴ the ratio is 19 : 5.

2. **9 : 7**

$$\begin{aligned}\text{No. choc-chip} &= 8 - 4\tfrac{1}{2}\\ &= 3\tfrac{1}{2}\\ \text{Ratio} &= 4\tfrac{1}{2} : 3\tfrac{1}{2}\\ &= 9 : 7\end{aligned}$$

∴ the ratio is 9 : 7.

3. **8 : 7**

$$\begin{aligned}\text{New price} &= 64\,000 - 8000\\ &= 56\,000\\ \text{Ratio} &= 64\,000 : 56\,000\\ &= 8 : 7\end{aligned}$$

∴ the ratio is 8 : 7.

4. **13 : 12**

$$\begin{aligned}\text{Jogging distance} &= 3 - (0.5 + 1.2)\\ &= 1.3\\ \text{Ratio} &= 1.3 : 1.2\\ &= 13 : 12\end{aligned}$$

∴ the ratio is 13 : 12.

5. **19 : 31**

$$\begin{aligned}\text{Percentage unread} &= 100 - (42 + 20)\\ &= 38\\ \text{Ratio} &= 38 : 62\\ &= 19 : 31\end{aligned}$$

∴ the ratio is 19 : 31.

6. **17 : 7**

$$\begin{aligned}\text{Students on excursion} &= \frac{1}{4} \times 28\\ &= 7\\ \text{Helping principal} &= 0.15 \times (28 - 7 - 1)\\ &= 0.15 \times 20\\ &= 3\\ \text{Present in class} &= 28 - 7 - 1 - 3 = 17\\ \text{Ratio} &= 17 : 7\end{aligned}$$

∴ the ratio is 17 : 7.

Key Skill  15 Ratio and rates: Simplifying ratios B

(pages 42–43)

1. **6 : 4 : 5**

$$\begin{aligned}\text{red : blue} &= 3 : 2\\ &= 6 : 4 \text{ (mult. by 2)}\\ \text{blue : black} &= 4 : 5\\ \text{red : blue : black} &= 6 : 4 : 5\end{aligned}$$

∴ the ratio is 6 : 4 : 5.

2. **9 : 4**

$$\begin{aligned}\text{coffee : ice-choc} &= 15 : 2\\ &= 45 : 6 \text{ (mult. by 3)}\\ \text{ice-choc : tea} &= 3 : 10\\ &= 6 : 20 \text{ (mult. by 2)}\\ \text{coffee : tea} &= 45 : 20\\ &= 9 : 4\end{aligned}$$

∴ the ratio is 9 : 4.

3. **9 : 6 : 10**

$$\begin{aligned}\text{red : blue} &= 3 : 2\\ &= 9 : 6 \text{ (mult. by 3)}\\ \text{blue : black} &= 3 : 5\\ &= 6 : 10 \text{ (mult. by 2)}\\ \text{red : blue : black} &= 9 : 6 : 10\end{aligned}$$

∴ the ratio is 9 : 6 : 10.

4. **4 : 3**

$$\begin{aligned}\text{wins : draws} &= 3 : 2\\ &= 6 : 4 \text{ (mult. by 2)}\\ \text{wins : losses} &= 2 : 1\\ &= 6 : 3 \text{ (mult. by 2)}\\ \text{draws : losses} &= 4 : 3\end{aligned}$$

∴ the ratio is 4 : 3.

5. **300**

Theo : Ronni = 5 : 3
= 10 : 6 (mult. by 2)
Ronni : Ted = 6 : 5
Theo : Ronni : Ted = 10 : 6 : 5

If Theo has 600 stamps, then Ted has 300 stamps.

∴ Ted has 300 stamps.

6. **3 : 5**

green : red = 2 : 1
= 6 : 3 (mult. by 2)
red : blue = 3 : 4
blue : yellow = 2 : 5
= 4 : 10 (mult. by 2)

∴ green : red : blue : yellow = 6 : 3 : 4 : 10

∴ green : yellow = 6 : 10
= 3 : 5

∴ the ratio is 3 : 5.

Key Skill  16 Ratio and rates: Dividing quantities in a given ratio (pages 44–45)

1. **$10 000**

Ratio of contributions = 6 : 4
= 3 : 2
Total parts = 3 + 2
= 5

$$\text{Sienna's amount} = \frac{2}{5} \times 25\,000 = 10\,000$$

∴ Sienna's share is $10 000.

2. **10 kg, 15 kg, 20 kg**

Total parts = 2 + 3 + 4
= 9

$$\text{Nitrates} = \frac{2}{9} \times 45 = 10$$

$$\text{Potash} = \frac{3}{9} \times 45 = 15$$

$$\text{Phosphates} = \frac{4}{9} \times 45 = 20$$

∴ 10 kg nitrates, 15 kg potash and 20 kg phosphates are in a 45-kg bag.

3. **$3500 and $4900**

Ratio of investments = 30 : 42
= 5 : 7
Total parts = 5 + 7
= 12

$$\text{First woman} = \frac{5}{12} \times 8400 = 3500$$

Second woman = 8400 − 3500
= 4900

∴ the women receive $3500 and $4900.

4. **9 cm, 15 cm, 12 cm**

Total parts = 3 + 5 + 4
= 12

$$\text{Side 1} = \frac{3}{12} \times 36 = 9$$

$$\text{Side 2} = \frac{5}{12} \times 36 = 15$$

Side 3 = 36 − (9 + 15)
= 12

∴ the sides are 9 cm, 15 cm and 12 cm.

5. **192 cm^2**

Length + width = 28
Total parts = 4 + 3
= 7

$$\text{Length} = \frac{4}{7} \times 28 = 16$$

Width = 28 − 16
= 12
Area = 16 × 12
= 192

∴ the area is 192 cm^2.

6. **6000, 3600, 6300**

A : B = 5 : 3
= 20 : 12 (mult. by 4)
B : C = 4 : 7
= 12 : 21 (mult. by 3)
A : B : C = 20 : 12 : 21
Total parts = 20 + 12 + 21
= 53

$$\text{Pop. of A} = \frac{20}{53} \times 15\,900 = 6000$$

$$\text{Pop. of B} = \frac{12}{53} \times 15\,900 = 3600$$

Pop. of C = 15 900 − (6000 + 3600)
= 6300

∴ the populations are 6000, 3600 and 6300.

Key Skill  17 Ratio and rates: Ratio involving unitary method (pages 46–47)

1. **36 cm**

5 parts = 60
3 parts = 60 ÷ 5 × 3
= 36

∴ the shorter side is 36 cm.

2. **240 kg and 160 kg**

$5 \text{ parts} = 400$
$3 \text{ parts} = 400 \div 5 \times 3$
$= 240$
$2 \text{ parts} = 400 \div 5 \times 2$
$= 160$

$\therefore$ 240 kg sand and 160 kg cement would be required.

3. **\$1760, \$1100, \$3520**

$3 \text{ parts} = 660$
$8 \text{ parts} = 660 \div 3 \times 8$
$= 1760$
$5 \text{ parts} = 660 \div 3 \times 5$
$= 1100$
$\text{Total} = 1760 + 1100 + 660$
$= 3520$

$\therefore$ other shares were \$1760 and \$1100, and the total was \$3520.

4. **13.5 cm**

$3 \text{ parts} = 8.1$
$5 \text{ parts} = 8.1 \div 3 \times 5$
$= 13.5$

$\therefore$ the longest side is 13.5 cm.

5. **\$52, \$117**

$9 - 4 = 5$

$\therefore$ the profit is 5 parts.

$5 \text{ parts} = 65$
$4 \text{ parts} = 65 \div 5 \times 4$
$= 52$
$9 \text{ parts} = 65 \div 5 \times 9$
$= 117$

$\therefore$ the cost price was \$52 and the selling price was \$117.

6. **8 mL**

$3 \text{ parts} = 120$
$1 \text{ parts} = 120 \div 3$
$= 40$

$\therefore$ 40 mL in Mrs Bradbury's solution

$\text{Total parts} = 1 + 4$
$= 5$
$\text{Initial acid} = \frac{1}{5} \times 40$
$= 8$

$\therefore$ Mrs Bradbury used 8 mL of acid.

Key Skill  18 Ratio and rates: Map scale (pages 48–49)

1. **4.8 cm**

$\text{Land distance in cm} = 48 \times 100\,000$
$= 4\,800\,000$

$\therefore$ the towns are 4 800 000 cm apart

$\text{Map distance} = 4\,800\,000 \div 1\,000\,000$
$= 4.8$

$\therefore$ the towns are 4.8 cm apart on the map.

2. **65 km**

$\text{Land distance in cm} = 26 \times 250\,000$
$= 6\,500\,000$

$\therefore$ the lookouts are 6 500 000 cm apart.

$\text{Land distance in km} = 6\,500\,000 \div 100\,000$
$= 65$

$\therefore$ the lookouts are 65 km apart.

3. **1 : 400 000, 72 km**

$120 \times 100\,000 = 12\,000\,000$

$\therefore$ the towns are 12 000 000 cm apart.

$\text{Scale} = 30 : 12\,000\,000$
$= 1 : 400\,000$
$\text{Land distance in cm} = 18 \times 400\,000$
$= 7\,200\,000$

$\therefore$ the villages are 7 200 000 cm apart.

$\text{Land distance in km} = 7\,200\,000 \div 100\,000$
$= 72$

$\therefore$ the villages are 72 km apart.

4. **3.4 km^2**

$\text{Length} = 8.5 \times 25\,000 = 212\,500$

$\therefore$ the length is 212 500 cm.

$\text{Length in km} = 212\,500 \div 100\,000$
$= 2.125 \quad \therefore 2.125 \text{ km.}$
$\text{Width} = 6.4 \times 25\,000 = 160\,000$

$\therefore$ the width is 160 000 cm.

$\text{Width in km} = 160\,000 \div 100\,000$
$= 1.6 \quad \therefore 1.6 \text{ km.}$
$\text{Area} = 2.125 \times 1.6$
$= 3.4$

$\therefore$ the area is 3.4 km^2.

5. **16 cm**

Map A: $\text{Land distance in cm} = 24 \times 8000$
$= 192\,000$

$\therefore$ the towns are 192 000 cm apart.

Map B: $\text{Map distance in cm} = 192\,000 \div 12\,000$
$= 16$

$\therefore$ the towns are 16 cm apart on Map B.

[Alternatively, the towns are closer on Map B with the smaller scale (1 : 12 000). As $\frac{8000}{12\,000} = \frac{2}{3}$, then find $\frac{2}{3}$ of $24 = 16$.]

6. **2.5 ha**

Suppose the park is rectangular with sides 25 cm by 10 cm.

$\text{Land length in cm} = 25 \times 1000$
$= 25\,000$

$\therefore$ length of the park is 25 000 cm, or 250 m.

$\text{Land width in cm} = 10 \times 1000$
$= 10\,000$

$\therefore$ length of the park is 10 000 cm, or 100 m.

$\text{Area of park} = 250 \times 100$
$= 25\,000$

$\therefore$ area of the park is 25 000 m^2, or 2.5 ha.

Revision Test 5
Level of difficulty—Average (page 50)

1. **72°**
   Other 2 angles add to 120°.
   $$\text{Total parts} = 3 + 2 = 5$$
   $$\text{Largest angle} = \frac{3}{5} \times 120 = 72$$
   ∴ the largest angle is 72°.

2. **5 : 1**
   $$\text{Fraction read} = \frac{1}{3} + \frac{1}{2} = \frac{5}{6}$$
   $$\text{Read} : \text{unread} = \frac{5}{6} : \frac{1}{6} = 5 : 1$$
   ∴ the ratio is 5 : 1.

3. **60 g and 150 g**
   4 parts = 120
   1 part = 30
   2 parts = 60
   5 parts = 150
   ∴ there is 60 grams of rice and 150 grams of oats.

4. **20**
   Tom : Mia = 4 : 3
   = 8 : 6 (mult. by 2)
   Mia : Aiden = 2 : 5
   = 6 : 15 (mult. by 3)
   Tom : Mia : Aiden = 8 : 6 : 15
   15 parts = 30
   1 part = 2
   8 parts = 16
   ∴ Tom is 16 years old now and will be 20 in 4 years' time.

5. **4.25 km**
   Land distance in cm = 17 × 25 000
   = 425 000
   ∴ the towers are 425 000 cm apart.
   Land distance in km = 425 000 ÷ 100 000
   = 4.25
   ∴ the towers are 4.25 km apart.

6. **6 : 5 : 7**
   Let Jack's height = 1
   ∴ Ben is 1.2 and Ethan is 1.4.
   ∴ Ratios = 1.2 : 1 : 1.4
   = 12 : 10 : 14
   = 6 : 5 : 7
   ∴ the ratio is 6 : 5 : 7.

Revision Test 6
Level of difficulty—Challenging (page 51)

1. **675 cm^2**
   Perimeter = 8 parts
   Sum of length and width = 4 parts
   ∴ width = 1 part
   8 parts = 120
   1 part = 15
   3 parts = 45
   ∴ the length is 45 cm and width is 15 cm.
   Area = 45 × 15
   = 675
   ∴ the area is 675 cm^2.

2. **40°**
   Total parts = 4 + 3 + 2
   = 9
   9 parts = 180
   1 part = 20
   4 parts = 80
   2 parts = 40
   80 − 40 = 40
   ∴ the difference is 40°.

3. **\$150, \$300**
   Difference in parts = 5 − 3
   = 2
   2 parts = 150
   4 parts = 300
   ∴ Len receives \$150 and Jen receives \$300.

4. **80**
   Sophia : Ava = 12 : 7
   Sophia : Ella = 3 : 4
   = 12 : 16
   Sophia : Ava : Ella = 12 : 7 : 16
   Total parts = 35
   35 parts = 175
   1 part = 5
   16 parts = 80
   ∴ Ella received 80 marks.

5. **8 : 12 : 18 : 27**
   A : B = 2 : 3
   = 8 : 12
   B : C = 2 : 3
   = 12 : 18
   C : D = 2 : 3
   = 18 : 27
   ∴ A : B : C : D = 8 : 12 : 18 : 27
   ∴ the ratio is 8 : 12 : 18 : 27.

6. **250 cm²**

Let the park be rectangular with sides 100 km by 10 km.

Park length in cm = 100 × 100 000
= 10 000 000

Map: Park length in cm = 10 000 000 ÷ 200 000
= 50

Park width in cm = 10 × 100 000
= 1 000 000

Map: Park length in cm = 1 000 000 ÷ 200 000
= 5

Map: Area = 50 × 5
= 250

∴ area is 250 cm².

Key Skill  Ratio and rates: Rates A (pages 52–53)

1. **$41**

15 bags of 4 lemons = 2.8 × 15
= 42

10 bags of 6 lemons = 4.1 × 10
= 41

6 bags of 10 lemons = 6.9 × 6
= 41.4

∴ best deal is 6-lemon bag.

∴ total cost is $41.

2. **8.75 h**

Bree: Hours worked = 500 ÷ 12.5
= 40

Lara: Hours worked = 500 ÷ 16
= 31.25

40 − 31.25 = 8.75

∴ it will take Bree an extra 8.75 h.

3. **$16.80**

5 packets of 6 rolls = 3.6 × 5
= 18

3 packets of 8 rolls and 1 packet of 6 rolls
= 4.4 × 3 + 3.6
= 16.8

2 packets of 12 rolls and 1 packet of 8 rolls
= 6.9 × 2 + 4.4
= 18.2

∴ best deal is 3 packets of 8 rolls and 1 packet of 6 rolls.

∴ the total cost is $16.80.

4. **63c**

95-octane:

Price per 100 km = 179.9 × 55 ÷ 580 × 100
= 1706 (nearest whole)

∴ price is $17.06/100 km.

91-octane:

Price per 100 km = 168.9 × 55 ÷ 525 × 100
= 1769 (nearest whole)

∴ price is $17.69/100 km

∴ the 91-octane is 63 cents/100 km more expensive.

5. **$0.15**

12 20-screw packs = 4.2 × 12
= 50.4

2 100-screw packs and 2 20-screw packs
= 13.5 × 2 + 4.2 × 2
= 35.4

∴ cheapest to buy 2 100-screw packs and 2 20-screw packs.

Average = 35.4 ÷ 240
= 0.1475
= 0.15 (2 dec. pl.)

∴ the cheapest cost is $0.15 per screw.

6. **No, Sid will need to increase savings by $11.25.**

Sam: 600 + 155 × 16 = 3080

Sid: 900 + 120 × 16 = 2820

∴ Sam will have enough money but Sid will not.

New weekly savings = (3000 − 900) ÷ 16
= 131.25

131.25 − 120 = 11.25

∴ Sid has to increase his weekly savings by $11.25.

Key Skill  Ratio and rates: Rates B (pages 54–55)

1. **$1797.45**

Rewrite 0.521 cents as $0.005 21.

Rates = 345 000 × 0.005 21
= 1797.45

∴ rates will be $1797.45.

2. **43.74 L**

540 ÷ 100 = 5.4

Petrol = 8.1 × 5.4
= 43.74

∴ Quentin used 43.74 litres.

3. **$320 000**

Jack: Rate calculation = 1476 ÷ 300 000
= 0.004 92

Set rate at 4.92 cents per dollar

Jill: Land value = (1476 + 98.4) ÷ 0.004 92
= 320 000

∴ Jill's land is valued at $320 000.

4. **$51.69**

380 ÷ 100 = 3.8

Petrol = 8.3 × 3.8
= 31.54

∴ car used 31.54 litres

Cost = 1.639 × 31.54
= 51.69 406
= 51.69 (2 dec. pl.)

∴ the cost will be $51.69.

5. 9000 km

Litres used $= 1431.90 \div 1.85$
$= 774$
$\therefore$ Kristie used 774 L.
Distance $= 774 \div 8.6 \times 100$
$= 9000$
$\therefore$ Kristie drove 9000 km.

6. $165.04

Cost per kL $= 167.85 \div 90$
$= 1.865$
New price $= 1.865 \times 1.035$
$= 1.930\,275$
$\therefore$ new price is $1.930 275 per kL.
New usage $= 90 \times 0.95$
$= 85.5$
$\therefore$ Ahmed has used 85.5 kL.
Cost $= 1.930\,275 \times 85.5$
$= 165.038\,5125 \ldots$
$= 165.04$ (2 dec. pl.)
$\therefore$ Ahmed will pay $165.04.

Key Skill  Ratio and rates: Distance, speed and time A (pages 56–57)

1. 273 km

Elapsed time = 11:45 am to 3:15 pm
= 3 h 30 min
= 3.5 h
Distance $= 78 \times 3.5$
$= 273$
$\therefore$ the motorist travelled 273 km.

2. 225 km

First distance $= 20 \times 5$
$= 100$
Second distance $= 25 \times 5$
$= 125$
$100 + 125 = 225$
$\therefore$ the total distance was 225 km.

3. 1800 km

Use 3 h as 180 min.
Number of 10 min $= 180 \div 10$
$= 18$
Distance $= 100 \times 18$
$= 1800$
$\therefore$ the plane would travel 1800 km.

4. 17.5 km

Elapsed time = 8:30 am to midday
= 3.5 h
Distance $= 3.5 \times 5$
$= 17.5$
$\therefore$ the cars are 17.5 km apart at midday.

5. 337.5 km

First distance $= 70 \times 3$
$= 210$
Second distance $= 85 \times 1.5$
$= 127.5$
$210 + 127.5 = 337.5$
$\therefore$ Kaitlyn travelled 337.5 km.

6. 24 ha

Change 10 min to $\frac{1}{6}$ h
Distance $= 12 \times \frac{1}{6}$
$= 2$
$\therefore$ the perimeter is 2 km.
Length + width = 1
$3 + 2 = 5$ parts
$\therefore$ total parts is 5
Length $= \frac{3}{5} \times 1$
$= 0.6$
$\therefore$ the length of the park is 600 m.
$\therefore$ the width of the park is 400 m.
Area $= 600 \times 400$
$= 240\,000$
Now, $240\,000 \div 10\,000 = 24$.
$\therefore$ the area of the park is 24 ha.

Key Skill  Ratio and rates: Distance, speed and time B (pages 58–59)

1. 86 km/h

Distance $= 64\,559 - 63\,871$
$= 688$
Average speed $= 688 \div 8$
$= 86$
$\therefore$ Sara averaged 86 km/h.

2. 60 km/h

Elapsed time = 4:45 pm to 6:15 pm
= 1.5 h
Average speed $= 90 \div 1.5$
$= 60$
$\therefore$ the bus averaged 60 km/h.

3. 75 km/h

Train 2: Average speed $= 360 \div 4$
$= 90$
6 parts $= 90$
5 parts $= 90 \div 6 \times 5$
$= 75$
$\therefore$ the average speed of the first train is 75 km/h.

4. **4 km/h**
Normal:
Elapsed time = 7:40 am to 9 am
= 1 h 20 min
= $1\frac{1}{3}$ h
Average speed = $48 \div 1\frac{1}{3}$
= 36
∴ Alice usually averages 36 km/h.
This morning:
Time = 1 h 30 min
= $1\frac{1}{2}$ h
Average speed = $48 \div 1\frac{1}{2}$
= 32
∴ she averaged 32 km/h.
36 − 32 = 4
∴ the difference was 4 km/h.

5. **700 km/h, 100 km/h**
Speed with wind = 2400 ÷ 3
= 800
Speed against wind = 2400 ÷ 4
= 600
Average of 2 speeds = 700
The plane flies 700 km/h in still air.
As 800 − 700 = 100, the speed of wind is 100 km/h.
∴ the plane flies at 700 km/h and the wind is 100 km/h.

6. **82 km/h**
Time for first half = 150 ÷ 75
= 2
Time for second half = 150 ÷ 90
= $1\frac{2}{3}$
Average speed = $300 \div 3\frac{2}{3}$
= 81.818 1818 …
= 82 (nearest whole)
∴ the truckdriver averaged 82 km/h.

Key Skill  Ratio and rates: Distance, speed and time C (pages 60–61)

1. **5:50 pm**
Time = 32 ÷ 64
= 0.5
∴ the journey took 30 min.
5:20 pm plus 30 min = 5:50 pm
∴ Liam arrived at 5:50 pm.

2. **9:55 pm**
Time = 851 ÷ 92
= 9.25
∴ the journey took 9 h 15 min
12:40 pm plus 9 h 15 min = 21:55 = 9:55 pm
∴ Grace arrived at 9:55 pm.

3. **10:45 am**
Time = 88 ÷ 32
= 2.75
∴ journey took 2 h 45 min.
1:30 pm minus 2 h 45 min
= 11:30 am minus 45 min
= 10:45 am
∴ she started at 10:45 am.

4. **4 h 16 min**
Time = 478 ÷ 112
= 4.267 857 143 …
= 4 h 16 min 4.3 s
= 4 h 16 min (nearest min)
∴ it will take 4 h 16 min between bends.

5. **Pete**
Larry's time = 48 ÷ 6
= 8
6 am plus 8 h = 2 pm
∴ Larry arrives at 2 pm.
Pete's time = 48 ÷ 16
= 3
10:45 am plus 3 h = 1:45 pm
∴ Pete arrives at 1:45 pm.
∴ Pete arrives first.

6. **3 min**
Length (distance) = 2 + 1
= 3
Time = 3 ÷ 60
= 0.05
= 3 min
∴ it will take 3 min.

Key Skill  Ratio and rates: Distance, speed and time D (pages 62–63)

1. **500 m**
Difference in speed = 8 − 6
= 2
∴ difference in speed is 2 km/h.
As 15 min = $\frac{1}{4}$ h,
Distance = $2 \times \frac{1}{4}$
= $\frac{1}{2}$
$\frac{1}{2}$ km = 500 m
∴ the distance is 500 m.

2. **4:40 pm**
Difference in speed $= 92 - 80$
$= 12$
$\therefore$ difference in speed is 12 km/h.
Time $= 20 \div 12$
$= 1\frac{2}{3}$
$1\frac{2}{3}$ h $=$ 1 h 40 min
3 pm plus 1 h 40 min $=$ 4:40 pm
$\therefore$ the cars are 20 km apart at 4:40 pm.

3. **4 km**
Difference in speed $= 38 - 32$
$= 6$
$\therefore$ difference in speed is 6 km/h.
Elapsed time $=$ 8:55 pm to 9:35 pm
$=$ 40 min
As 40 min $= \frac{2}{3}$ h,
Distance $= 6 \times \frac{2}{3}$
$= 4$
$\therefore$ Mr Alder is 4 km from home.

4. **45 km**
Difference in speed $= 80 - 50$
$= 30$
$\therefore$ difference in speed is 30 km/h.
As 1 h 30 min $= 1\frac{1}{2}$ h,
Distance $= 30 \times 1\frac{1}{2}$
$= 45$
$\therefore$ the cyclist still has 45 km to travel when the motorist arrives.

5. **400 m**
Difference in speed $= 100 - 64$
$= 36$
$\therefore$ difference in speed is 36 km/h.
As $36 \times 1000 \div 60 \div 60 = 10$,
then 36 km/h $=$ 10 m/s.
Length (distance) $= 10 \times 40$
$= 400$
$\therefore$ the train is 400 m long.

6. **5:42 pm**
Matthew: Time $= 270 \div 90$
$= 3$
4:30 pm minus 3 h $=$ 1:30 pm
$\therefore$ they left Smithfield at 1:30 pm.
Total distance $= 270 + 45$
$= 315$
Blake: Time $= 315 \div 75$
$= 4.2$
$=$ 4 h 12 min
1:30 pm plus 4 h 12 min $=$ 5:42 pm
$\therefore$ Blake will arrive at 5:42 pm.

Key Skill  Ratio and rates: Distance, speed and time E

(pages 64–65)

1. **3250 km**
Relative speed $= 720 + 780$
$= 1500$
Elapsed time $=$ 7:10 am to 9:20 am
$=$ 2 h 10 min
$= 2\frac{1}{6}$ h
Distance $= 1500 \times 2\frac{1}{6}$
$= 3250$
$\therefore$ the planes are 3250 km apart.

2. **10:10 am**
Relative speed $= 7 + 5$
$= 12$
Time $= 16 \div 12$
$= 1\frac{1}{3}$
$=$ 1 h 20 min
8:50 am plus 1 h 20 min $=$ 10:10
$\therefore$ they meet at 10:10 am.

3. **7:47:30 am**
Relative speed $= 8 + 8$
$= 16$
Time $= 2 \div 16$
$= 0.125$
$=$ 7 min 30 s
7:40 am plus 7 min 30 s $=$ 7:47:30
$\therefore$ the girls are 2 km apart at 7:47:30 am.

4. **8:20 pm, 160 km from Eastburg**
Relative speed $= 80 + 90$
$= 170$
Time $= 340 \div 170$
$= 2$
6:20 pm plus 2 h $=$ 8:20 pm
$\therefore$ the trains meet at 8:20 pm.
Train A: Distance $= 80 \times 2$
$= 160$
$\therefore$ Train A has travelled 160 km.
$\therefore$ the trains meet at 8:20 pm, 160 km from Eastburg.

5. **3 km**
Relative speed $= 4 + 6$
$= 10$
As 18 min $= \frac{18}{60} = \frac{3}{10}$ h,
Distance $= 10 \times \frac{3}{10}$
$= 3$
$\therefore$ the length is 3 km.

6. **5:00 pm**

$$\text{Relative speed} = 6 + 9 = 15$$

$$\text{As 40 min} = \frac{2}{3}\text{ h,}$$

$$\text{Distance apart} = 15 \times \frac{2}{3} = 10$$

∴ they are 10 km apart at 4:10 pm.

$$\text{Jarrod: Time} = 10 \div 12 = \frac{5}{6} = 50\text{ min}$$

4:10 pm plus 50 min = 5:00 pm

∴ they meet again at 5:00 pm.

Revision Test 7
Level of difficulty—Average (page 66))

1. **$237 255**

Nathan: Rate calculation = 1530 ÷ 220 000
= 0.006 954 545 …
= 0.06c in dollar

1530 + 120 = 1650

Lauren: Land value = 1650 ÷ 0.0069 …
= 237 254.902 …
= 237 255
(nearest whole)

∴ Lauren's property is worth $237 255.

2. **18 m**

$72 \times 1000 \div 60 \times 60 = 20$

∴ 72 km/h = 20 m/s

Distance = 20×0.9
= 18

∴ the bridge is 18 m wide.

3. **9:22 pm**

Time A = $84 \div 70$
= 1.2

∴ Zoe travelled for 1 h 12 min.

$$\text{Time B} = 120 \div 90 = 1\frac{1}{3}$$

∴ Zoe travelled for 1 h 20 min.

Total time = 1 h 12 m + 20 m + 1 h 20 m
= 2 h 52 min

6:30 pm plus 2 h 52 min = 9:22 pm

∴ Zoe reached her destination at 9:22 pm.

4. **36 km**

4 pm to 9 pm is 5 h.

Train A: Distance = 52×5
= 260

Train B: Distance = 64×3.5
= 224

Difference = 260 − 224
= 36

∴ the drivers are 36 km apart.

5. **4 km**

To school: 6 km/h means 12 km in 2 h.
From school: 12 km/h means 12 km in 1 h.
Total: 24 km in 3 h
Average speed = 8 km/h
If the total trip takes 1 h, then the total distance is 8 km, or from home to school is 4 km.

6. **15 km/h**

$$1 - \left(\frac{1}{5} + \frac{1}{3}\right) = \frac{7}{15}$$

Seven-fifteenths of trip = 28
Fifteen-fifteenths of trip = 28 ÷ 7 × 15
= 60

∴ the distance is 60 km

Average speed = 60 ÷ 4
= 15

∴ Mel's average speed was 15 km/h.

Revision Test 8
Level of difficulty—Challenging (page 67)

1. **Karen**

31 − 12 = 19

∴ there are 19 days left in the month.

Karen: Total pages = 320 + 20 × 19
= 700

Lesley: Total pages = 380 + 16 × 19
= 684

∴ only Karen will finish her book.

2. **$128.40**

Last year: Rate calculation = 1200 ÷ 340 000
= 0.003 529 …

New rate calculation = 0.003 529 … × 1.025
= 0.003 617 647 …

New land value = 340 000 × 1.08
= 367 200

New rates cost = 367 200 × 0.003 617 …
= 1328.4

1328.4 − 1200 = 128.4

∴ rates increased by $128.40.

3. **2 min 45.6 s**

Length = 2.5 + 0.95
= 3.45

Time = 3.45 ÷ 75
= 0.046
= 2 min 45.6 s

∴ it will take 2 min 45.6 s.

4. **2 h 57 min**
$\frac{1}{3} \times 54 = 18$
$\therefore$ each third is 18 km.
Time for first third $= 18 \div 24$
$= 0.75$
Time for second third $= 18 \div 18$
$= 1$
Time for final third $= 18 \div 15$
$= 1.2$
Total time $= 0.75 + 1 + 1.2$
$= 2.95$
$=$ 2 h 57 min
$\therefore$ the cyclist took 2 h 57 min.

5. **45 min**
Relative speed $= 6 + 9$
$= 15$
Distance $= 15 \times 0.5$
$= 7.5$
$\therefore$ entire circuit is 7.5 km.
Angela's time $= 7.5 \div 6$
$= 1.25$
Time required $= 1.25 - 0.5$
$= 0.75$
$=$ 45 min
$\therefore$ Angela takes another 45 min.

6. **3 tonne**
Perimeter $= 11.4 \times \frac{28}{60}$
$= 5.32$
As $5.32 \times 1000 = 5320$, the perimeter is 5320 m.
Length + width $= 2660$
As $5 + 3 = 8$,
Length $= \frac{5}{8} \times 2660$
$= 1662.5$
Width $= \frac{3}{8} \times 2.66$
$= 997.5$
Dimensions are 1662.5 m by 997.5 m.
Area $= 1662.5 \times 997.5$
$= 1\,658\,343.75$
As $1\,658\,343.75 \div 10\,000 = 165.834\,375$, then the area is 165.834 375 ha.
Mass $= 20 \times 165.834\,375$
$= 3316.6875$
Amount in tonne $= 3316.6875 \div 1000$
$= 3.316\,6875$
$\therefore$ council uses 3 tonne (nearest tonne).

ALGEBRA

Key Skill  26 Equations A (pages 68–69)

1. **$85**
Let the cost of the bangle be x.
$\therefore$ the cost of the watch is $x + 70$.
$x + x + 70 = 240$
$2x = 240 - 70$
$2x = 170$
$x = 85$
$\therefore$ the bangle costs $85.

2. **5.7 m, 2.9 m**
Let the length of the shorter section be x.
$\therefore$ the length of longer section is $x + 2.8$.
$x + x + 2.8 = 8.6$
$2x = 8.6 - 2.8$
$2x = 5.8$
$x = 2.9$
Also, $2.9 + 2.8 = 5.7$.
$\therefore$ the lengths are 5.7 m and 2.9 m.

3. **$785**
Let the cost of the computer be x.
$\therefore$ the cost of the television was $x + 220$.
$x + x + 220 = 1350$
$2x = 1350 - 220$
$2x = 1130$
$x = 565$
Also, $565 + 220 = 785$.
$\therefore$ the television cost $785.

4. **123 km**
Let the distance from Leighwood to Albert be x.
$\therefore$ the distance from Leighwood to Pompeda is $x + 80$.
$x + x + 80 = 326$
$2x = 326 - 80$
$2x = 246$
$x = 123$
$\therefore$ Leighwood is 123 km from Albert.

5. **$42.50**
Let Lia's contribution be x.
$\therefore$ Mia's contribution was $x + 35$.
$x + x + 35 = 120$
$2x = 120 - 35$
$2x = 85$
$x = 42.5$
$\therefore$ Lia contributed $42.50.

6. 20, 11, 8

Let the age of the middle boy be x.

$\therefore$ other boys are $x + 9$ and $x - 3$.

If the average is 13, then the total is 39.

$$x + x + 9 + x - 3 = 39$$
$$3x + 6 = 39$$
$$3x = 33$$
$$x = 11$$

Also $11 + 9 = 20$, $11 - 3 = 8$.

$\therefore$ the boys are 20, 11 and 8.

Key Skill  27 Equations B (pages 70–71)

1. 16

Let the number of rabbits be x.

$\therefore$ there are $2x$ cats and $4x$ dogs.

$$x + 2x + 4x = 28$$
$$7x = 28$$
$$x = 4$$

$\therefore$ number of dogs is $4 \times 4 = 16$.

$\therefore$ there are 16 dogs.

2. 40

Let Kim's age now be x.

$\therefore$ Ken's age now is $2x$.

10 years ago: Kim: $x - 10$

Ken: $2x - 10$

$$\therefore 2x - 10 = 3(x - 10)$$
$$2x - 10 = 3x - 30$$
$$3x - 2x = -10 + 30$$
$$x = 20$$

$\therefore$ Ken is $2 \times 20 = 40$.

$\therefore$ Ken is 40 years old.

3. 2

Let Shanais's age now be x.

$\therefore$ Serena is $3x$ and Ethan is $(x - 5)$.

6 years' time: Shanais: $x + 6$, Serena: $3x + 6$ and Ethan: $x + 1$

$$x + 6 + 3x + 6 + x + 1 = 48$$
$$5x + 13 = 48$$
$$5x = 48 - 13$$
$$5x = 35$$
$$x = 7$$

$\therefore$ Ethan is $7 - 5 = 2$.

$\therefore$ Ethan is 2 years old.

4. \$1.60

Let the cost of juice be x.

$\therefore$ cost of apple is $(x - 20)$, cost of wrap is $(x + 60)$.

$$x + x - 20 + x + 60 = 340$$
$$3x + 40 = 340$$
$$3x = 340 - 40$$
$$3x = 300$$
$$x = 100$$

$\therefore$ wrap is $100 + 60 = 160$.

$\therefore$ the wrap costs \$1.60.

5. 72

Let the number of children be x.

$\therefore$ the number of men is $3x$ and women is $6x$.

$$6x + 3x + x = 120$$
$$10x = 120$$
$$x = 12$$

$\therefore$ men is $3 \times 12 = 36$ and women is $2 \times 36 = 72$.

$\therefore$ there were 72 women at the reunion.

6. 14 and 56

Let Kate's age now be x.

$\therefore$ grandmother is now $4x$.

6 years' time: Kate: $x + 6$, grandmother: $4x + 6$

$$4x + 6 = 3(x + 6) + 2$$
$$4x + 6 = 3x + 18 + 2$$
$$4x + 6 = 3x + 20$$
$$4x - 3x = 20 - 6$$
$$x = 14$$

$\therefore$ grandmother is $14 \times 4 = 56$.

$\therefore$ Kate is 14 and her grandmother is 56.

Key Skill  28 Equations and consecutive numbers (pages 72–73)

1. 21, 22, 23

Let consecutive numbers be x, $x + 1$, $x + 2$.

$$x + x + 1 + x + 2 = 66$$
$$3x + 3 = 66$$
$$3x = 66 - 3$$
$$3x = 63$$
$$x = 21$$

$\therefore x + 1 = 22$ and $x + 2 = 23$

$\therefore$ the numbers are 21, 22 and 23.

2. 13, 15, 17

Let consecutive odd numbers be x, $x + 2$, $x + 4$.

$$x + x + 2 + x + 4 = 45$$
$$3x + 6 = 45$$
$$3x = 45 - 6$$
$$3x = 39$$
$$x = 13$$

$\therefore x + 2 = 15$ and $x + 4 = 17$

$\therefore$ the numbers are 13, 15 and 17.

3. −9, −8, −7

Let consecutive numbers be x, $x + 1$, $x + 2$

$$x + x + 1 + x + 2 = -24$$
$$3x + 3 = -24$$
$$3x = -24 - 3$$
$$3x = -27$$
$$x = -9$$

$\therefore x + 1 = -8$ and $x + 2 = -7$

$\therefore$ the numbers are -9, -8 and -7.

4. **15, 16, 17**

Let consecutive numbers be x, $x + 1$, $x + 2$.

$x + x + 2 = 32$
$2x + 2 = 32$
$2x = 32 - 2$
$2x = 30$
$x = 15$

$\therefore x + 1 = 16$ and $x + 2 = 17$

$\therefore$ the numbers are 15, 16 and 17.

5. **2, 3, 4**

Let consecutive numbers be x, $x + 1$, $x + 2$.

$4(x + x + 1) = 5(x + 2)$
$4(2x + 1) = 5(x + 2)$
$8x + 4 = 5x + 10$
$8x - 5x = 10 - 4$
$3x = 6$
$x = 2$

$\therefore x + 1 = 3$ and $x + 2 = 4$

$\therefore$ the numbers are 2, 3 and 4.

6. **1, 3, 5**

Let consecutive odd numbers be x, $x + 2$, $x + 4$.

$5(x + 2) = 2(x + x + 4) + 3$
$5(x + 2) = 2(2x + 4) + 3$
$5x + 10 = 4x + 8 + 3$
$5x + 10 = 4x + 11$
$5x - 4x = 11 - 10$
$x = 1$

$\therefore x + 2 = 3$ and $x + 4 = 5$

$\therefore$ the numbers are 1, 3 and 5.

Key Skill  29 Equations and measurement

(pages 74–75)

1. **27 cm by 9 cm**

Let width be x.

$\therefore$ the length is $3x$.

Also, length + breath = $72 \div 2$
$= 36$

$x + 3x = 36$
$4x = 36$
$x = 9$

$\therefore 3x = 3 \times 9 = 27$

$\therefore$ the rectangle is 27 cm by 9 cm.

2. **1536 m^2**

Let length be x.

$\therefore$ the width is $x - 16$.

Also, length + breath = $160 \div 2$
$= 80$

$x + x - 16 = 80$
$2x - 16 = 80$
$2x = 80 + 16$
$2x = 96$
$x = 48$

$\therefore x - 16 = 48 - 16 = 32$

$\therefore$ the rectangle is 48 m by 32 m.

Area $= 48 \times 32$
$= 1536$

$\therefore$ the area is 1536 m^2.

3. **38 m by 12 m**

Let width be x.

$\therefore$ the length is $3x + 2$.

Also, length + breath = $100 \div 2$
$= 50$

$x + 3x + 2 = 50$
$4x + 2 = 50$
$4x = 50 - 2$
$4x = 48$
$x = 12$

$\therefore 3x + 2 = 3 \times 12 + 2 = 38$

$\therefore$ the rectangle is 38 m by 12 m.

4. **115 m by 65 m**

Perimeter $= 1116 \div 3.1$
$= 360$

Also, length + breath = $360 \div 2$
$= 180$

Let width be x.

$\therefore$ the length is $x + 50$.

$x + x + 50 = 180$
$2x + 50 = 180$
$2x = 180 - 50$
$2x = 130$
$x = 65$

$\therefore x + 50 = 115$

$\therefore$ the dimensions are 115 m by 65 m.

5. **15 m, 30 m, 21 m**

Perimeter $= 155.1 \div 2.35$
$= 66$

Let the shortest side be x.

$\therefore$ the other sides are $2x$ and $x + 6$.

$x + 2x + x + 6 = 66$
$4x + 6 = 66$
$4x = 66 - 6$
$4x = 60$
$x = 15$

$\therefore 2x = 2 \times 15 = 30$, $x + 6 = 15 + 6 = 21$

$\therefore$ the sides are 15 m, 30 m and 21 m.

6. **576 cm^2**

Let each side of the square be x.

$\therefore$ the length of each side of the triangle is $x + 8$.

$4x = 3(x + 8)$
$4x = 3x + 24$
$4x - 3x = 24$
$x = 24$

$\therefore$ the square has side 24 cm.

Area $= 24^2$
$= 576$

$\therefore$ the area of the square is 576 cm^2.

Key Skill  30 Equations and speed

(pages 76–77)

1. **6 pm**

Let Layla's time of travel be x.

$\therefore$ Ava's time of travelling is $x - \frac{1}{6}$.

As Distance = Speed × Time:

$$45 \times x = 54(x - \frac{1}{6})$$
$$45x = 54x - 9$$
$$9x = 9$$
$$x = 1$$

5 pm plus 1 h = 6 pm

$\therefore$ Ava caught up at 6 pm.

2. **24 km**

Let Caleb's time of travel be x.

$\therefore$ Jax's time of travelling is $x - \frac{1}{3}$

As Distance = Speed × Time:

$$18 \times x = 24(x - \frac{1}{3})$$
$$18x = 24x - 8$$
$$6x = 8$$
$$x = 1\frac{1}{3}$$

$\therefore$ Jax caught up to Caleb after $1\frac{1}{3}$ h.

Caleb's distance $= 18 \times 1\frac{1}{3}$

$= 24$

$\therefore$ Caleb rode 24 km.

3. **$4\frac{1}{4}$ h**

Let first car's time of travel be x.

$\therefore$ second car's time of travelling is $x - \frac{3}{4}$.

As Distance = Speed × Time:

$$85 \times x = 100(x - \frac{3}{4})$$
$$85x = 100x - 75$$
$$15x = 75$$
$$x = 5$$

$$x - \frac{3}{4} = 5 - \frac{3}{4}$$
$$= 4\frac{1}{4}$$

$\therefore$ it takes $4\frac{1}{4}$ hours to overtake.

4. **dead-heat**

Let James' time of running be x.

$\therefore$ Ben's time of running is $x - \frac{1}{12}$.

As Distance = Speed × Time:

$$10 \times x = 12(x - \frac{1}{12})$$
$$10x = 12x - 1$$
$$2x = 1$$
$$x = \frac{1}{2}$$

$\therefore$ after half an hour the two athletes are together.

James's distance $= 10 \times \frac{1}{2}$

$= 5$

As 5 km = 5000 m, which is the length of the race, they cross the finish line at the same time.

$\therefore$ the race finished in a dead-heat.

5. **12 min**

Let Brian's time of running be x.

$\therefore$ Lucas's time of running is $x - \frac{1}{10}$

As Distance = Speed × Time:

$$8 \times x = 12(x - \frac{1}{10})$$
$$8x = 12x - \frac{6}{5}$$
$$4x = \frac{6}{5}$$
$$x = \frac{3}{10}$$

$\frac{3}{10} \times 60 = 18$

Now, $18 - 6 = 12$.

$\therefore$ Lucas passed Brian after 12 minutes of running.

6. **12:24 pm**

Let Jocelyn's time of travelling be x.

$\therefore$ Chloe's time of travelling is $x - 1$.

As Distance = Speed × Time:

$$90 \times x + 100(x - 1) = 356$$
$$90x + 100x - 100 = 356$$
$$190x = 456$$
$$x = 2.4$$
$$= 2 \text{ h } 24 \text{ min}$$

$\therefore$ at 12:24 pm they were 356 km apart.

Key Skill  31 Equations and mixtures

(pages 78–79)

1. **86**

Let the number of adult tickets be x.

$\therefore$ number of child's tickets be $(128 - x)$.

$$12 \times x + 8(128 - x) = 1368$$
$$12x + 1024 - 8x = 1368$$
$$4x + 1024 = 1368$$
$$4x = 344$$
$$x = 86$$

$\therefore$ 86 adult tickets were sold.

2. **30 kg**

Let the mass of millet seed be x.

$\therefore$ the mass of sunflower seed is $(50 - x)$.

$$1.8 \times x + 2.3(50 - x) = 50 \times 2$$
$$1.8x + 115 - 2.3x = 100$$
$$115 - 0.5x = 100$$
$$0.5x = 115 - 100$$

$$0.5x = 15$$
$$x = 30$$

$\therefore$ 30 kg of millet seed are needed.

3. **38 chickens and 28 goats**
   Let the number of chickens be x.
   $\therefore$ number of goats is $(66 - x)$
   As chickens have 2 legs and goats have 4 legs:
$$2 \times x + 4(66 - x) = 188$$
$$2x + 264 - 4x = 188$$
$$264 - 2x = 188$$
$$2x = 264 - 188$$
$$2x = 76$$
$$x = 38$$
   $\therefore 66 - x = 66 - 38 = 28$
   $\therefore$ 38 chickens and 28 goats were on the farm.

4. **1.6 kg of 30% and 2.4 kg of 80%**
   Let the mass of the 30% silver be x.
   $\therefore$ mass of the 80% silver is $4 - x$.
$$0.3 \times x + 0.8(4 - x) = 0.6 \times 4$$
$$0.3x + 3.2 - 0.8x = 2.4$$
$$3.2 - 0.5x = 2.4$$
$$0.5x = 0.8$$
$$x = 1.6$$
   $\therefore 4 - x = 4 - 1.6 = 2.4$
   $\therefore$ 1.6 kg of 30% silver and 2.4 kg of 80% silver will be needed.

5. **100**
   Let the number of 3-legged stools be x.
   $\therefore$ number of 4-legged stools is $(120 - x)$.
$$3 \times x + 4(120 - x) = 380$$
$$3x + 480 - 4x = 380$$
$$480 - x = 380$$
$$x = 480 - 380$$
$$x = 100$$
   $\therefore$ Leo made 100 3-legged stools.

6. **\$6500**
   Let the amount invested at 8% interest be x.
   $\therefore$ amount invested at 7% interest is $10\,000 - x$.
$$0.08 \times x + 0.07(10\,000 - x) = 765$$
$$0.08x + 700 - 0.07x = 765$$
$$0.01x + 700 = 765$$
$$0.01x = 765 - 700$$
$$0.01x = 65$$
$$x = 6500$$
   $\therefore$ \$6500 was invested at Sumbank.

Key Skill 32 Using formulae

(pages 80–81)

1. **\$3840**
$$I = Prn$$
$$= 12\,000 \times 0.04 \times 8$$
$$= 3840$$
   $\therefore$ the interest is \$3840.

2. **15 cm**
$$A = \frac{1}{2}bh$$
$$120 = \frac{1}{2} \times b \times 16$$
$$120 = 8b$$
$$b = \frac{120}{8}$$
$$= 15$$
   $\therefore$ the length is 15 cm.

3. **\$1440**
$$C = 480 + 120r$$
$$C = 480 + 120 \times 8$$
$$= 1440$$
   $\therefore$ it will cost \$1440.

4. **6 cm**
$$V = \frac{4}{3}\pi r^3$$
$$907.78 = \frac{4}{3}\pi r^3$$
$$r^3 = 907.78 \div \frac{4}{3}\pi$$
$$= 216.716\,5114 \ldots$$
$$r = 6.006\,627\,042 \ldots$$
$$= 6 \text{ (nearest whole)}$$
   $\therefore$ the radius is 6 cm.

5. **210 cm^2**
$$A = \frac{h}{2}(a + b)$$
$$= \frac{14}{2}(12 + 18)$$
$$= 7(30)$$
$$= 210$$
   $\therefore$ the area is 210 cm^2.

6. **95 kg**
$$\text{Previous BMI} = 28.4 + 2.26$$
$$= 30.66$$
$$I = \frac{m}{h^2}$$
$$30.66 = \frac{m}{h^2}$$
$$= \frac{m}{1.76^2}$$
$$m = 30.66 \times 1.76^2$$
$$= 94.972\,416 \ldots$$
$$= 95$$
   $\therefore$ Scott's mass was 95 kg.

Key Skill 33 Linear relationships

(pages 82–83)

1. **\$240**
$$C = 12n$$
$$= 12 \times 20$$
$$= 240$$
   $\therefore$ the cost is \$240.

2. **$55**
$C = 10t + 15$
$= 10 \times 4 + 15$
$= 55$
$\therefore$ the cost is $55.

3. **2**
$s = 50 - 3n$
$= 50 - 3 \times 16$
$= 2$
$\therefore$ Charlotte will have 2 soaps remaining.

4. **$71.98**
$C = 2.14d + 3.5$
$= 2.14 \times 32 + 3.5$
$= 71.98$
$\therefore$ it will cost $71.98.

5. **87.5 km**
$d = 800 - 95n$
$= 800 - 95 \times 7.5$
$= 87.5$
$\therefore$ Noah still has 87.5 km to drive.

6. **10 km**

n	1	2	4	10
D	3	3.5	4.5	7.5

$D = 2.5 + 0.5n$
$= 2.5 + 0.5 \times 15$
$= 10$
$\therefore$ Sophie ran 10 km.

Revision Test 9
Level of difficulty—Average (page 84)

1. **$595**
Let the cost of the tablet be x.
$\therefore$ the cost of the phone was $x + 200$.
$x + x + 200 = 990$
$2x + 200 = 990$
$2x = 990 - 200$
$2x = 790$
$x = 395$
$x + 200 = 395 + 200$
$= 595$
$\therefore$ the phone cost $595.

2. **21**
Let Jack's age today be x.
$\therefore$ Nigel's age today is $2x$.
Four years ago: Jack: $x - 4$ and Nigel: $2x - 4$
$2x - 4 = 3(x - 4)$
$2x - 4 = 3x - 12$
$3x - 2x = -4 + 12$
$x = 8$
$\therefore$ Jack is 8 and Nigel is 16.
$\therefore$ in 5 years, Nigel will be 21.

3. **13**
$20 - 3 = 17$
Let the number of correct answers be x.
$\therefore$ the number of incorrect answers is $17 - x$.
$5 \times x - 3(17 - x) = 53$
$5x - 51 + 3x = 53$
$8x = 53 + 51$
$8x = 104$
$x = 13$
$\therefore$ 13 answers were correct.

4. **904.78 cm²**
$S = 2\pi rh + 2\pi r^2$
$S = 2 \times \pi \times 8 \times 10 + 2 \times \pi \times 8^2$
$= 904.778\,6842\ldots$
$= 904.78$ (2 dec. pl.)
$\therefore$ the surface area is 904.78 cm^2.

5. **3:20 pm**
Let Maureen's travel time be x.
$\therefore$ Leif's travel time is $x - 2$.
As Distance = Speed × Time:
$80 \times x = 110(x - 2)$
$80x = 110x - 220$
$30x = 220$
$x = 7\frac{1}{3}$
$\therefore$ after $7\frac{1}{3}$h or 7 h 20 min
$\therefore$ Leif will catch Maureen at 3:20 pm.

6. **960 m²**
Let the width be x.
$\therefore$ the length is $2x + 8$.
$2(2x + 8 + x) = 136$
$2(3x + 8) = 136$
$6x + 16 = 136$
$6x = 136 - 16$
$6x = 120$
$x = 20$
$2x + 8 = 2 \times 20 + 8$
$= 48$
$\therefore$ the dimensions are 48 m by 20 m
Area $= 48 \times 20$
$= 960$
$\therefore$ the area is 960 m^2.

Revision Test 10
Level of difficulty—Challenging (page 85)

1. **13 and 68**
Let Theo's present age be x.
$\therefore$ Barry's age is $81 - x$.
Last year Theo was $x - 1$.
In four years, Barry will be $81 - x + 4 = 85 - x$.
$85 - x = 6(x - 1)$
$85 - x = 6x - 6$

$$6x + x = 85 + 6$$
$$7x = 91$$
$$x = 13$$
$$81 - x = 81 - 13$$
$$= 68$$

∴ Theo is 13 and Barry is 68.

2. **4, 6, 8**
   Let the numbers be x, $x + 2$, $x + 4$.
   $3(x + 2) = 3(x + x + 4) - 18$
   $3(x + 2) = 3(2x + 4) - 18$
   $3x + 6 = 6x + 12 - 18$
   $3x + 6 = 6x - 6$
   $6x - 3x = 6 + 6$
   $3x = 12$
   $x = 4$
   ∴ the numbers are 4, 6, 8.

3. **$800**

n	10	12	20	30
P	−100	0	400	900

   $P = 50n - 600$
   $P = 50 \times 28 - 600$
   $= 800$
   ∴ Sasha makes $800 profit.

4. **5**
   Let the number of sixes be x.
   ∴ let the number of fours be $3x$.
   $4 \times 3x + 6 \times x = 90$
   $12x + 6x = 90$
   $18x = 90$
   $x = 5$
   ∴ Adam hit 5 sixes.

5. **$112, $48**
   Let the amount of Libby's money be $7x$.
   ∴ let the amount of Grace's money be $3x$.
   $7x - 24 = 3x + 40$
   $7x - 3x = 40 + 24$
   $4x = 64$
   $x = 16$
   $7 \times x = 7 \times 16 = 112$
   $3 \times x = 3 \times 16 = 48$
   ∴ Libby had $112 and Grace $48.

6. **5:15 pm**
   Let the travel time of Carter be x.
   ∴ let the travel time of Alyssa be $x - \frac{1}{4}$.
   As Distance = Speed × Time:
   $80 \times x = 100(x - \frac{1}{4})$
   $80x = 100x - 25$
   $100x - 80x = 25$
   $20x = 25$
   $x = 1\frac{1}{4}$
   ∴ after $1\frac{1}{4}$ h
   Distance $= 80 \times 1\frac{1}{4}$
   $= 100$
   Carter and Alyssa are 200 km apart.
   As Alyssa travels at 100 km/h, she will reach Carter in 2 h.
   2 pm plus 1 h 15 min plus 2 h = 5:15 pm
   ∴ Alyssa reaches Carter at 5:15 pm.

MEASUREMENT

Key Skill 34 Circumference A

(pages 86–87)

1. **754 cm**
   $C = \pi d$
   $= \pi \times 30$
   $= 94.247\,779\,61 \ldots$
   $= 94.25$ (2 dec. pl.)
   Total length $= 94.25 \times 8$
   $= 753.982\,2369 \ldots$
   $= 754$ (nearest whole)
   ∴ the gold trim totals 754 cm.

2. **13 m**
   Inside lane:
   $C = 2\pi r$
   $= 2 \times \pi \times 55$
   $= 345.575\,1919 \ldots$
   $= 345.58$ (2 dec. pl.)
   Outside lane:
   $C = 2\pi r$
   $= 2 \times \pi \times 57$
   $= 358.121\,5625 \ldots$
   $= 358.12$ (2 dec. pl.)
   Difference $= 358.12 - 345.58$
   $= 12.54$
   $= 13$ (nearest whole)
   ∴ the runner has to run an extra distance of 13 m.

3. **126 mm**
   $C = 2\pi r$
   $= 2 \times \pi \times 8$
   $= 50.265\,482\,46 \ldots$
   $= 50.265$ (3 dec. pl.)
   Side length $= 50.265 \div 4$
   $= 12.566\,370\,61 \ldots$
   $= 12.6$ (1 dec. pl.)
   ∴ each side is 126 mm.

4. **1005 m**

$$\begin{aligned} C &= 2\pi r \\ &= 2 \times \pi \times 3.2 \\ &= 20.106\,192\,98 \ldots \\ &= 20.11 \text{ (2 dec. pl.)} \\ \text{Distance} &= 20.11 \times 50 \\ &= 1005.309\,649 \ldots \\ &= 1005 \text{ (nearest whole)} \end{aligned}$$

∴ the horse will run 1005 m.

5. **$179.97**

$$\begin{aligned} \text{Perimeter} &= \frac{1}{4} \times 2\pi r + 2r \\ &= \frac{1}{4} \times 2 \times \pi \times 8 + 2 \times 8 \\ &= 28.566\,370\,61 \ldots \\ &= 28.57 \text{ (2 dec. pl.)} \\ \text{Cost} &= 28.57 \times 6.3 \\ &= 179.968\,1349 \ldots \\ &= 179.97 \text{ (2 dec. pl.)} \end{aligned}$$

∴ edging costs $179.97.

6. **572 cm**

Hour hand: $C = 2\pi r$

$$\begin{aligned} &= 2 \times \pi \times 5 \\ &= 10\pi \end{aligned}$$

Hour hand makes 1 rotation.

$$\text{Distance} = 10\pi$$

Minute hand: $C = 2\pi r$

$$\begin{aligned} &= 2 \times \pi \times 8 \\ &= 16\pi \end{aligned}$$

Minute hand makes 12 rotations.

$$\begin{aligned} \text{Distance} &= 12 \times 16\pi \\ &= 192\pi \\ \text{Difference} &= 192\pi - 10\pi \\ &= 182\pi \\ &= 571.769\,863\ldots \\ &= 572 \text{ (nearest whole)} \end{aligned}$$

∴ the minute hand travels 572 cm further.

Key Skill 35 Circumference B

(pages 88–89)

1. **34 cm**

$$\begin{aligned} C &= 2\pi r \\ 68\pi &= 2\pi r \\ r &= \frac{68\pi}{2\pi} \\ &= 34 \end{aligned}$$

∴ the radius is 34 cm.

2. **14 cm**

$$\begin{aligned} \text{Perimeter} &= 2(12 + 10) \\ &= 44 \\ C &= \pi d \\ 44 &= \pi d \\ d &= \frac{44}{\pi} \\ &= 14.005\,634\,99 \ldots \\ &= 14 \text{ (nearest whole)} \end{aligned}$$

∴ the diameter is 14 cm.

3. **7 cm**

1.1 m = 110 cm

$$\begin{aligned} C &= 110 \div 5 \\ &= 22 \\ C &= \pi d \\ 22 &= \pi d \\ d &= \frac{22}{\pi} \\ d &= 7.002\,817\,496 \ldots \\ &= 7 \text{ (nearest whole)} \end{aligned}$$

∴ the diameter is 7 cm.

4. **11.37**

1 m = 1000 mm

$$\begin{aligned} C &= \pi d \\ &= \pi \times 28 \\ &= 87.964\,5943 \ldots \\ &= 87.96 \text{ (2 dec. pl.)} \\ \text{Revs} &= 1000 \div 87.96 \\ &= 11.368\,210\,22 \ldots \\ &= 11.37 \text{ (2 dec. pl.)} \end{aligned}$$

∴ the coin rolled 11.37 revolutions.

5. **32**

$$\begin{aligned} \text{Distance} &= \text{Speed} \times \text{Time} \\ &= 24 \times \frac{1}{3} \\ &= 8 \end{aligned}$$

∴ Jay rides 8 km, or 8000 m.

$$\begin{aligned} C &= \pi d \\ &= \pi \times 80 \\ &= 80\pi \\ \text{Revolutions} &= 8000 \div 80\pi \\ &= 31.830\,988\,62 \ldots \\ &= 32 \text{ (nearest whole)} \end{aligned}$$

∴ 32 revolutions of the track are completed.

6. **663**

$$\begin{aligned} \text{Front wheel } C &= 2\pi r \\ &= 2 \times \pi \times 0.72 \\ &= 4.523\,893\,421 \ldots \\ \text{No. of revolutions} &= 1000 \div 4.523\,893\,421\ldots \\ &= 221.048\,5321 \ldots \\ &= 221.05 \text{ (2 dec. pl.)} \end{aligned}$$

Rear wheel: radius is one-quarter the radius of front wheel.

$$\begin{aligned} \therefore \text{No. of revolutions} &= 221.05 \times 4 \\ &= 884.2 \\ \text{Difference} &= 884.2 - 221.05 \\ &= 663.15 \\ &= 663 \text{ (nearest whole)} \end{aligned}$$

∴ there were 663 more revolutions.

Key Skill  Area of quadrilaterals and triangles A

(pages 90–91)

1. 56 cm

$$\begin{aligned}\text{Area of square} &= 28^2\\ &= 784\\ \text{Area of rectangle} &= lb\\ 784 &= l \times 14\\ 14l &= 784\\ l &= 56\end{aligned}$$

$\therefore$ length is 56 cm.

2. 8 cm

$$\begin{aligned}\text{Area of kite} &= \frac{1}{2} \times 16 \times 12\\ &= 96\\ \text{Area of triangle} &= \frac{1}{2}bh\\ 96 &= \frac{1}{2} \times 24 \times h\\ 12h &= 96\\ h &= 8\end{aligned}$$

$\therefore$ the height is 8 cm.

3. $\sqrt{12}$ cm

$$\begin{aligned}\text{Area of square} &= 6^2\\ &= 36\end{aligned}$$

Let the height of the parallelogram be x.
The base of the parallelogram is $3x$.
Area of parallelogram $= bh$
$36 = 3x \times x$
$3x^2 = 36$
$x^2 = 12$
$x = \sqrt{12}$
$\therefore$ the height is $\sqrt{12}$ cm.

4. 208 m

$$\begin{aligned}\text{Area of rectangle} &= 64 \times 30\\ &= 1920\end{aligned}$$

$\therefore$ area is 1920 m^2

$$\begin{aligned}\text{Cost of fertilising} &= 96 \div 1920\\ &= 0.05\end{aligned}$$

$\therefore$ cost of fertiliser is \$0.05/m^2

$$\begin{aligned}\text{Area of square} &= 135.2 \div 0.05\\ &= 2704\\ \text{Side of square} &= \sqrt{2704}\\ &= 52\end{aligned}$$

$\therefore$ square has side 52 m.

$$\begin{aligned}\text{Perimeter of square} &= 52 \times 4\\ &= 208\end{aligned}$$

$\therefore$ the perimeter of the first paddock is 208 m.

5. \$302.40

$$\begin{aligned}\text{Area of base} &= 6 \times 3\\ &= 18\\ \text{Area of walls} &= 2 \times 6 \times 2 + 2 \times 3 \times 2\\ &= 36\\ \text{Total area} &= 18 + 36\\ &= 54\\ \text{Total cost} &= 54 \times 5.6\\ &= 302.4\end{aligned}$$

$\therefore$ the cost of painting is \$302.40.

6. 1 : 4

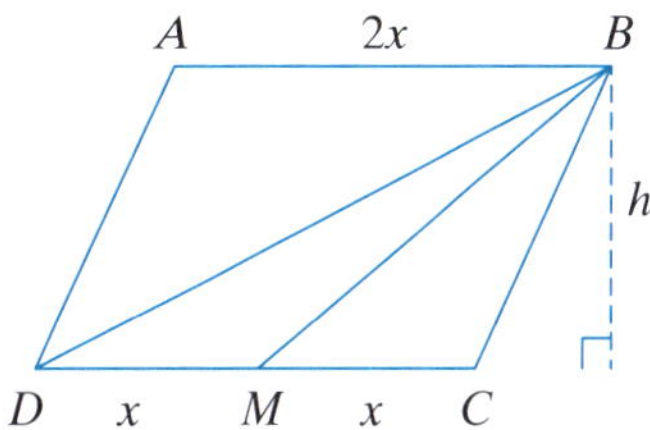

Let AB be $2x$, then DM is x.
Also, let the height of the parallelogram/triangle be h.

$$\begin{aligned}\text{Area of triangle} &= \frac{1}{2} \times x \times h\\ &= \frac{xh}{2}\\ \text{Area of parallelogram} &= 2x \times h\\ &= 2xh\\ \text{Ratio} &= \frac{xh}{2} : 2xh\\ &= xh : 4xh\\ &= 1 : 4\end{aligned}$$

$\therefore$ the ratio is 1 : 4.

Key Skill  Area of quadrilaterals and triangles B

(pages 92–93)

1. 8 cm

$$\begin{aligned}\text{Area of rhombus} &= \frac{1}{2} \times 20 \times 8\\ &= 80\\ \text{Area of parallelogram} &= bh\\ 80 &= 10 \times h\\ 10h &= 80\\ h &= 8\end{aligned}$$

$\therefore$ the height is 8 cm.

2. 18 cm

$$\begin{aligned}\text{Area of rhombus} &= \frac{1}{2} \times 15 \times 12\\ &= 90\\ \text{Area of trapezium} &= \frac{1}{2}h(a + b)\\ 90 &= \frac{1}{2} \times 10 \times (a + b)\\ 5(a + b) &= 90\\ a + b &= 18\end{aligned}$$

$\therefore$ the sum is 18 cm.

3. **10 cm**

$$\text{Area of kite} = \frac{1}{2} \times 15 \times 8$$
$$= 60$$
$$\text{Area of triangle} = \frac{1}{2}bh$$
$$60 = \frac{1}{2} \times b \times 12$$
$$6b = 60$$
$$b = 10$$

$\therefore$ the base is 10 cm.

4. **12 m**

$$\text{Total area of both shapes} = 350.4 \div 3.65$$
$$= 96$$

$\therefore$ each shape has area of 48 m^2

$$\text{Area of parallelogram} = bh$$
$$48 = b \times 4$$
$$4b = 48$$
$$b = 12$$

$\therefore$ the base is 12 m.

5. **9 cm**

$$\text{Perimeter} = 4s$$
$$4s = 24$$
$$s = 6$$
$$\text{Area of square} = 6^2$$
$$= 36$$
$$\text{Area of kite} = \frac{1}{2}xy$$
$$36 = \frac{1}{2} \times 8 \times y$$
$$4y = 36$$
$$y = 9$$

$\therefore$ the other diagonal is 9 cm.

6. **300 m**

2.4 t = 2400 kg

$$\text{Area} = 2400 \div 160$$
$$= 15$$

$\therefore$ the area of each park is 15 ha, or 150 000 m^2

$$\text{Area of rhombus} = \frac{1}{2}xy$$
$$150\,000 = \frac{1}{2} \times 1000 \times y$$
$$500y = 150\,000$$
$$y = 300$$

$\therefore$ the other diagonal is 300 m long.

Key Skill 38 Area of circles A

(pages 94–95)

1. **32.170 m²**

$$A = \pi r^2$$
$$= \pi \times 3.2^2$$
$$= 32.169\,908\,77\ldots$$
$$= 32.170 \text{ (3 dec. pl.)}$$

$\therefore$ the area of grass is 32.170 m^2.

2. **386 cm²**

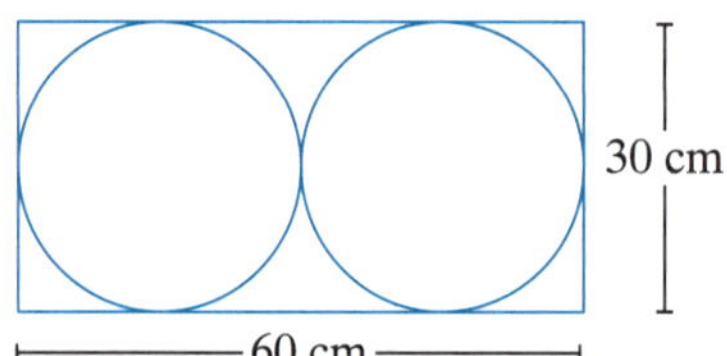

$$\text{Remaining area} = 60 \times 30 - 2 \times \pi \times 15^2$$
$$= 386.283\,3059\ldots$$
$$= 386 \text{ (nearest whole)}$$

$\therefore$ the area remaining is 386 cm^2.

3. **5 m²**

$$\text{Diameter} + 2 \text{ overhangs} = 2.3 + 2 \times 0.1$$
$$= 2.5$$

$\therefore$ radius is 1.25 m

$$\text{Area} = \pi \times 1.25^2$$
$$= 4.908\,738\,521\ldots$$
$$= 5 \text{ (nearest whole)}$$

$\therefore$ the area of the tablecloth is 5 m^2.

4. **4.05 cm²**

$$\text{Area} = \pi \times 2.3^2 - \pi \times 2^2$$
$$= 4.052\,654\,523\ldots$$
$$= 4.05 \text{ (2 dec. pl.)}$$

$\therefore$ the area of metal is 4.05 cm^2.

5. **50.27 units²**

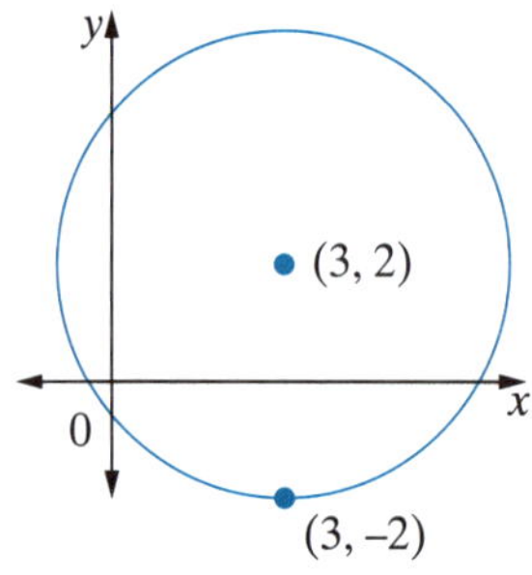

Circle has radius of 4 units.

$$\text{Area} = \pi \times 4^2$$
$$= 50.265\,482\,46\ldots$$
$$= 50.27 \text{ (2 dec. pl.)}$$

$\therefore$ the area of the circle is 50.27 $units^2$.

6. **80π m²**

$$\text{Shaded area} = \text{Area of large semicircle} + \text{area of small semicircle}$$
$$A = \frac{1}{2} \times \pi \times 12^2 + \frac{1}{2} \times \pi \times 4^2$$
$$= 72\pi + 8\pi$$
$$= 80\pi$$

$\therefore$ the shaded area is 80π m^2.

Key Skill 39 Area of circles B

(pages 96–97)

1. **75.40 cm^2**

Shaded area = Area of large semicircle − area of small semicircle

$$A = \frac{1}{2} \times \pi \times 8^2 - \frac{1}{2} \times \pi \times 4^2$$
$$= 75.398\,223\,69\ldots$$
$$= 75.40$$

$\therefore$ the shaded area is 75.40 cm^2.

2. **34.27 cm^2**

Shaded area = Area of quadrant − area of triangle

$$A = \frac{1}{4} \times \pi \times 8^2 - \frac{1}{2} \times 8 \times 4$$
$$= 34.265\,482\,46\ldots$$
$$= 34.27$$

$\therefore$ the shaded area is 34.27 cm^2.

3. **4π cm^2**

Shaded area = Area of large quadrant − area of small quadrant

$$A = \frac{1}{4} \times \pi \times 5^2 - \frac{1}{4} \times \pi \times 3^2$$
$$= \frac{1}{4}(25\pi - 9\pi)$$
$$= 4\pi$$

$\therefore$ the shaded area is 4π cm^2.

4. **13.73 cm^2**

$$\text{Side} = \frac{32}{4}$$
$$= 8$$

$\therefore$ the diameter is 8 cm

Shaded area = Area of square − area of circle

$$A = 8^2 - \pi \times 4^2$$
$$= 13.734\,517\,54\ldots$$
$$= 13.73 \text{ (2 dec. pl.)}$$

$\therefore$ the shaded area is 13.73 cm^2

5. **$\frac{3\pi}{2}$ cm^2**

Radius = 3 cm and $\angle NOM = 60°$

$$\text{Area} = \frac{1}{6} \times \pi \times 3^2$$
$$= \frac{3\pi}{2}$$

$\therefore$ the shaded area is $\frac{3\pi}{2}$ cm^2.

Revision Test 
Level of difficulty—Average

(page 98)

1. **2 653 000**

$$\text{Circumference} = \pi d$$
$$= 60\pi$$

As $5000 \times 1000 \times 100 = 500\,000\,000$, then the journey is 500 000 000 cm.

$$\text{Number} = 500\,000\,000 \div 60\pi$$
$$= 2\,652\,823.85\ldots$$
$$= 2\,653\,000 \text{ (nearest thousand)}$$

$\therefore$ tyre rotates 2 653 000 times.

2. **16 cm**

$$\text{Area of rhombus} = \frac{1}{2} \times 24 \times 16$$
$$= 192$$
$$\text{Area of parallelogram} = bh$$
$$192 = b \times 12$$
$$12b = 192$$
$$b = 16$$

$\therefore$ the base is 16 cm.

3. **14.4 cm**

$$\text{Side of square} = 48 \div 4$$
$$= 12$$
$$\text{Area of square} = 12^2$$
$$= 144$$

$\therefore$ area of trapezium is 144 cm^2

$$\text{Area of trapezium} = \frac{1}{2}h(a + b)$$
$$144 = \frac{1}{2}h(20)$$
$$10h = 144$$
$$h = 14.4$$

$\therefore$ the perpendicular height is 14.4 cm.

4. **1.93 units^2**

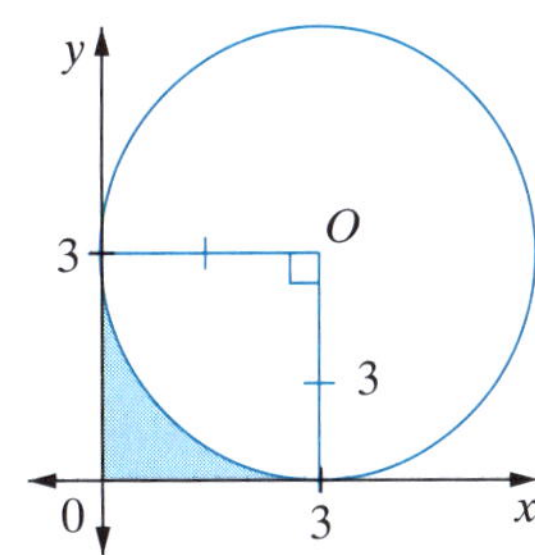

Shaded area = Area of square − area of quadrant

$$= 3^2 - \frac{1}{4} \times \pi \times 3^2$$
$$= 1.931\,416\,529\ldots$$
$$= 1.93 \text{ (2 dec. pl.)}$$

$\therefore$ the area of the shaded region is 1.93 units^2.

5. **35 m**

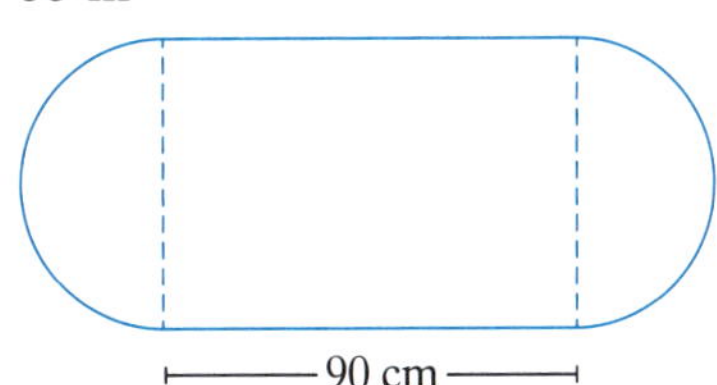

$$\text{Perimeter} = 2\pi r + 2 \times 90$$
$$400 = 2\pi r + 2 \times 90$$
$$400 = 2\pi r + 180$$

$$2\pi r = 220$$
$$r = \frac{220}{2\pi}$$
$$r = 35.014\,087\,48\ldots$$
$$= 35 \text{ (nearest whole)}$$
$\therefore$ radius is 35 m.

6. 21%

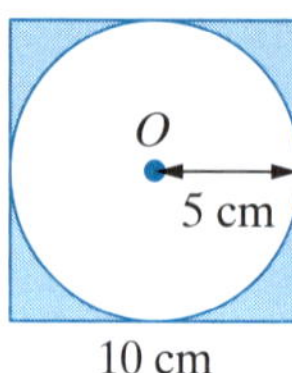

$$\text{Square side} = 40 \div 4$$
$$= 10$$
$$\text{Area of square} = 10^2$$
$$= 100$$
$\therefore$ area is 100 cm^2

$$\text{Shaded area} = \text{Area of square} - \text{area of circle}$$
$$= 100 - \pi \times 5^2$$
$$= 21.460\,183\,66\ldots$$
$$= 21 \text{ (nearest whole)}$$
$$\text{Percentage shaded} = \frac{21}{100} \times 100\%$$
$$= 21\%$$
$\therefore$ 21% of the square is not covered.

Revision Test 12
Level of difficulty—Challenging (page 99)

1. 35

$$\text{Length of rope} = 50 \times 1.3$$
$$= 65 \text{ m}$$
$$\therefore \text{Circumference} = \pi d$$
$$= \pi \times 0.6$$
$$= 1.884\,955\,592\ldots$$
$$\text{No. of revolutions} = 65 \div 1.884\,955\,592\ldots$$
$$= 34.483\,571\ldots$$
$$= 35 \text{ (complete revs.)}$$
$\therefore$ will need 35 revolutions (34 will not bring the bucket to the top of the well).

2. 215 L

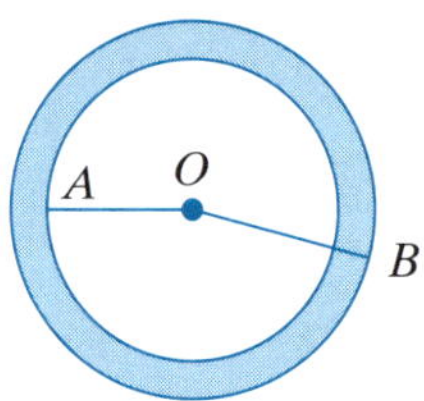

$$\text{Shaded area} = \text{Area of large circle} - \text{area of small circle}$$
$$= \pi \times 14^2 - \pi \times 10^2$$
$$= 96\pi$$
$\therefore$ area is 96π m^2

$$\text{Number of litres} = 96\pi \div 1.4$$
$$= 215.423\,4962\ldots$$
$$= 215 \text{ (nearest whole)}$$
$\therefore$ need 215 L.

3. $(4\pi + 20)$ cm

$$\text{Area of semicircle} = \frac{1}{2}\pi r^2$$
$$8\pi = \frac{1}{2}\pi r^2$$
$$\frac{1}{2}r^2 = 8$$
$$r^2 = 16$$
$$r = 4 \quad (r > 0)$$
$\therefore SR = l = 8$ cm

$$\text{Perimeter of rectangle} = 2(l + b)$$
$$28 = 2(8 + b)$$
$$14 = b + 8$$
$$b = 6$$
Perimeter of shaded region
$$= \frac{1}{2} \times \pi \times 8 + 8 + 6 + 6$$
$$= 4\pi + 20$$
$\therefore$ the perimeter is $(4\pi + 20)$ cm.

4. $(8 + \pi)$ cm

Let the length of the rectangle be x.
$$\text{Total area} = \text{Area of rectangle} + \text{area of semicircle}$$
$$= 2x + \frac{1}{2}\pi r^2$$
$$6 + \frac{1}{2}\pi = 2x + \frac{1}{2} \times \pi \times 1^2$$
$$6 + \frac{1}{2}\pi = 2x + \frac{1}{2}\pi$$
$$\therefore 2x = 6$$
$$x = 3$$
$\therefore$ the length of the rectangle is 3 cm.
$$\text{Perimeter} = \frac{1}{2} \times 2\pi \times 1 + 2 + 3 + 3$$
$$= 8 + \pi$$
$\therefore$ the perimeter is $(8 + \pi)$ cm.

5. 177

$$\text{Circumference} = \pi d$$
$$= 60\pi$$
$$\text{Pedal rotation} = 60\pi \times 3$$
$$= 180\pi$$
$\therefore$ the bike travels 180π cm every rotation of the pedals.

As 1 km = 100 000 cm:
$$\text{No. of pedal rotations} = 100\,000 \div 180\pi$$
$$= 176.838\,8257\ldots$$
$\therefore$ needs to be 177 rotations of the pedals.

6. 6

Todd: $\quad \text{Race time} = \text{Distance} \div \text{Speed}$
$$= 20 \div 40$$
$$= 0.5$$
$\therefore$ Todd finishes the race in 30 min.

Ken: $\quad \text{Distance} = \text{Speed} \times \text{Time}$
$$= 36 \times 0.5$$
$$= 18$$

As $20 - 18 = 2$, Todd still has 2 km, or 2000 m to travel.

$$\begin{aligned}\text{Track circumference} &= 2\pi r\\ &= 2 \times \pi \times 60\\ &= 120\pi\\ \text{Number of laps} &= 2000 \div 120\pi\\ &= 5.305\,167\,77\ldots\\ &= 6 \text{ (to finish the race)}\end{aligned}$$

$\therefore$ Ken still has 6 laps to ride.

Key Skill 40 Volume of prisms

(pages 100–101)

1. **26 cm**

$$\begin{aligned}\text{Vol. of rectangular prism} &= lbh\\ &= 52 \times 26 \times 13\\ &= 17\,576\end{aligned}$$

$\therefore$ volume is 17 576 cm^3

$$\begin{aligned}\text{Volume of cube} &= s^3\\ s^3 &= 17\,576\\ s &= \sqrt[3]{17\,576}\\ &= 26\end{aligned}$$

$\therefore$ the sides are 26 cm.

2. **16 cm**

$$\begin{aligned}\text{Vol. of rectangular prism} &= lbh\\ 6912 &= 24 \times 18 \times h\\ 432h &= 6912\\ h &= 16\end{aligned}$$

$\therefore$ the height is 16 cm.

3. **36.75 cm**

$$\begin{aligned}\text{Volume of cube} &= 21^3\\ &= 9261\end{aligned}$$

$\therefore$ volume is 9261 cm^3

$$\begin{aligned}\text{Vol. of rectangular prism} &= lbh\\ 9261 &= 18 \times 14 \times h\\ 252h &= 9261\\ h &= 36.75\end{aligned}$$

$\therefore$ the height is 36.75 cm.

4. **5400 m^3**

$$\begin{aligned}27\text{ ha} &= 270\,000\text{ m}^2\\ 20\text{ mm} &= 0.02\text{ m}\\ \text{Volume} &= 270\,000 \times 0.02\\ &= 5400\end{aligned}$$

$\therefore$ 5400 m^3 of water fell.

5. **652.5 mm^3**

$$\begin{aligned}\text{Area} &= \frac{1}{2}h(a + b)\\ &= \frac{1}{2} \times 15 \times (11 + 18)\\ &= 217.5\end{aligned}$$

$\therefore$ the area is 217.5 mm^2

$$\begin{aligned}\text{Volume} &= 217.5 \times 3\\ &= 652.5\end{aligned}$$

$\therefore$ the volume of the medallion is 652.5 mm^3.

6. **2 cm**

$$\begin{aligned}\text{Area of choc. bar} &= \frac{1}{2}bh\\ &= \frac{1}{2} \times 3 \times 3\\ &= 4.5\end{aligned}$$

$\therefore$ the area is 4.5 cm^2

Let depth of choc. bar be x.

$\therefore$ Volume of choc. bar $= 4.5x$

$$\begin{aligned}\text{Volume of block} &= 20 \times 20 \times 9\\ &= 3600\\ \therefore \quad 400 \times 4.5x &= 3600\\ 1800x &= 3600\\ x &= 2\end{aligned}$$

$\therefore$ the depth of chocolate is 2 cm.

Key Skill  Capacity

(pages 102–103)

1. **30 cm**

$72\text{ L} = 72\,000\text{ cm}^3$

$$\begin{aligned}\text{Vol. of rectangular prism} &= lbh\\ 72\,000 &= 60 \times 40 \times h\\ 2400h &= 72\,000\\ h &= 30\end{aligned}$$

$\therefore$ height of tank is 30 cm.

2. **1 m 12 s**

$80\text{ mL} = 80\text{ cm}^3$

$$\begin{aligned}V &= Ah\\ &= 480 \times 12\\ &= 5760\\ \text{Time} &= 5760 \div 80\\ &= 72\end{aligned}$$

72 s = 1 min 12 s

$\therefore$ it will take 1 m 12 s.

3. **6 cm**

$$\begin{aligned}\text{One-quarter of container} &= 54\\ \text{Four-quarters of container} &= 54 \times 4\\ &= 216\end{aligned}$$

$\therefore$ container can hold 216 mL, or 216 cm^3

$$\begin{aligned}\text{Side} &= \sqrt[3]{216}\\ &= 6\end{aligned}$$

$\therefore$ each side is 6 cm.

4. **2.24 L**

$$\begin{aligned}\text{Vol. required} &= 20 \times 16 \times 7\\ &= 2240\end{aligned}$$

$2240\text{ cm}^3 = 2240$ mL or 2.24 L

$\therefore$ the container needs another 2.24 L of water.

5. **16 cm**

$388.8\text{ L} = 388\,800\text{ cm}^3$

$$\begin{aligned}V &= \frac{1}{2}h(a + b) \times 1800\\ 388\,800 &= \frac{1}{2}h(12 + 15) \times 1800\\ 24\,300h &= 388\,800\\ h &= 16\end{aligned}$$

$\therefore$ the depth of the gutter is 16 cm.

6. 80%

$$4 \text{ parts} = 6.4$$
$$1 \text{ part} = 6.4 \div 4$$
$$= 1.6$$

$6.4 + 1.6 = 8$

∴ Chris uses 8 L of cordial.

$8 \text{ L} = 8000 \text{ cm}^3$

$$V = Ah$$
$$\text{Total volume} = 500 \times 20$$
$$= 10\,000$$

Total capacity is 10 L.

As Chris used 8 L, the percentage is 80%.

∴ the dispenser is 80% full.

Key Skill  42 Volume of cylinders

(pages 104–105)

1. A by 239 cm^3

$$\text{Cylinder A: } V = \pi r^2 h$$
$$= \pi \times 6^2 \times 16$$
$$= 576\pi$$
$$\text{Cylinder B: } V = \pi r^2 h$$
$$= \pi \times 5^2 \times 20$$
$$= 500\pi$$
$$\text{DIfference} = 576\pi - 500\pi$$
$$= 76\pi$$
$$= 239 \text{ (nearest whole)}$$

∴ Cylinder A has the greater volume by 239 cm^3.

2. 19 min

$$V = \pi \times 6^2 \times 10$$
$$= 1130.973\,355\ldots$$
$$= 1131 \text{ (nearest whole)}$$

∴ capacity is 1131 mL.

$$1131 \div 60 = 18.85$$
$$= 19 \text{ (nearest whole)}$$

∴ it will take 19 min.

3. 44 cm

$$V = \pi r^2 h$$
$$20\,000 = \pi \times 12^2 \times h$$
$$h = \frac{20\,000}{144\pi}$$
$$= 44.209\,706\,41\ldots$$
$$= 44 \text{ (nearest whole)}$$

∴ it is 44 cm tall.

4. 13.03 cm

$$V = \pi r^2 h$$
$$1600 = \pi \times r^2 \times 12$$
$$r^2 = \frac{1600}{12\pi}$$
$$r = \sqrt{\frac{1600}{12\pi}}$$
$$= 6.514\,700\,159\ldots$$
$$d = 13.029\,400\,32\ldots$$
$$= 13.03 \text{ (2 dec. pl.)}$$

∴ the diameter is 13.03 cm.

5. 7238 L

$$V = \frac{3}{4}\pi r^2 h$$
$$= \frac{3}{4} \times \pi \times 1.6^2 \times 1.2$$
$$= 7.238\,229\,474\ldots$$
$$= 7.238 \text{ (3 dec. pl.)}$$

Family needs 7.238 kL, or 7238 L.

∴ 7238 L of water should be used.

6. 7 kg

External radius = 5 cm

Internal radius = 4.5 cm

$$V = \pi \times 5^2 \times 45 - \pi \times 4.5^2 \times 45$$
$$= 671.515\,4297\ldots$$
$$= 671.52 \text{ (2 dec. pl.)}$$
$$\text{Mass} = 671.52 \times 11$$
$$= 7386.669\,727\ldots$$
$$= 7386 \text{ (nearest whole)}$$

7386 g = 7.386 kg, or 7 kg (nearest kg)

∴ the pipe has a mass of 7 kg.

Key Skill  43 Pythagoras' theorem A

(pages 106–107)

1. 5 km

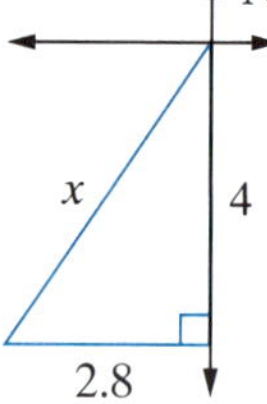

$$x^2 = 4^2 + 2.8^2$$
$$= 23.84$$
$$x = \sqrt{23.84}$$
$$= 4.882\,622\,46\ldots$$
$$= 5 \text{ (nearest whole)}$$

∴ the third leg is 5 km.

2. 4.8 m

$$x^2 = 8^2 - 6.4^2$$
$$= 23.04$$
$$x = \sqrt{23.04}$$
$$= 4.8$$

∴ the base is 4.8 m from pole.

3. 17 cm

$$x^2 = 15^2 + 8^2$$
$$= 289$$
$$x = \sqrt{289}$$
$$= 17$$

∴ the longest straw is 17 cm.

4. 71%

Let the square have sides 1 unit.

$$x^2 = 1^2 + 1^2$$
$$= 2$$
$$x = \sqrt{2}$$
$$= 1.414\,213\,562\ldots$$
$$= 1.414 \text{ (3 dec. pl.)}$$

∴ Rose walked 2 km, Andrew walked 1.414 km.

$$\text{Percentage} = \frac{1.414}{2} \times 100\%$$
$$= 70.710\,678\,12\ldots$$
$$= 71 \text{ (nearest whole)}$$

∴ Andrew walked 71% of the distance Rose walked.

5. **40 cm**

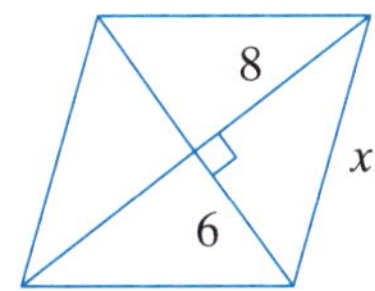

$$\text{Area} = \frac{1}{2}\text{product of diagonals}$$
$$96 = \frac{1}{2} \times 12 \times d$$
$$6d = 96$$
$$d = 16$$

∴ diagonals are 12 and 16 cm.

$$x^2 = 8^2 + 6^2$$
$$= 100$$
$$x = 10$$

∴ side length is 10 cm

$$10 \times 4 = 40$$

∴ the perimeter is 40 cm.

6. **14 m 21 s**

$$x^2 = 18^2 + 15^2$$
$$= 549$$
$$x = \sqrt{549}$$
$$x = 23.430\,749\,03\ldots$$
$$= 23.43 \text{ (2 dec. pl.)}$$
$$\text{Difference} = (18 + 15) - 23.43$$
$$= 9.57$$

∴ the direct route is 9.57 km shorter.

$$\text{Time} = \frac{9.57}{40}$$
$$= 0.239\,25$$
$$= 14 \text{ min } 21 \text{ s}$$

∴ Lachlan would save 14 min 21 s.

Key Skill 44 Pythagoras' theorem B (pages 108–109)

1. **$155**

$$\text{Area} = \frac{1}{2}bh$$
$$80 = \frac{1}{2} \times 16 \times h$$
$$8h = 80$$
$$h = 10$$

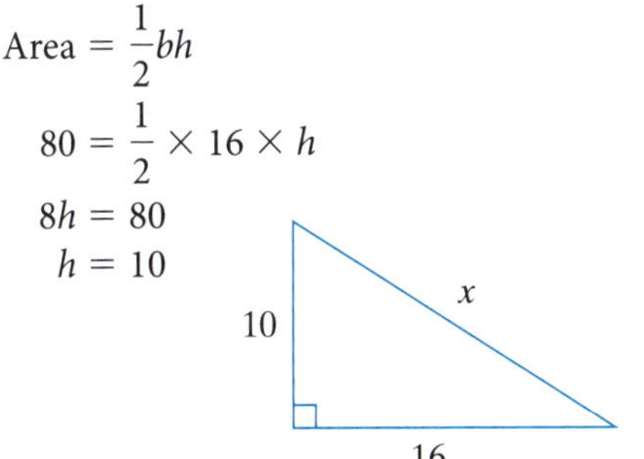

$$x^2 = 10^2 + 16^2$$
$$= 356$$
$$x = \sqrt{356}$$
$$= 18.867\,962\,26\ldots$$
$$= 18.87 \text{ (2 dec. pl.)}$$
$$\text{Perimeter} = 10 + 16 + 18.87$$
$$= 44.87$$
$$\text{Cost} = 44.87 \times 3.45$$
$$= 154.8015$$
$$= 155 \text{ (nearest whole)}$$

∴ the cost is $155.

2. **$20.48**

$$32^2 = x^2 + x^2$$
$$2x^2 = 1024$$
$$x^2 = 512$$

∴ area is 512 m^2

$$\text{Cost} = 512 \times 0.04$$
$$= 20.48$$

∴ the cost is $20.48.

3. **$(2\pi - 4)$ cm^2**

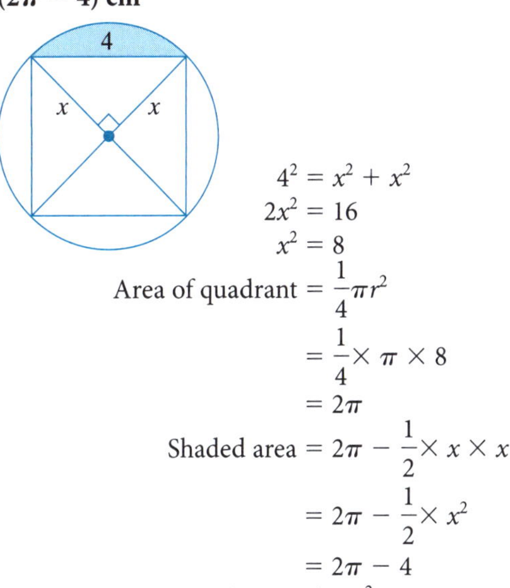

$$4^2 = x^2 + x^2$$
$$2x^2 = 16$$
$$x^2 = 8$$
$$\text{Area of quadrant} = \frac{1}{4}\pi r^2$$
$$= \frac{1}{4} \times \pi \times 8$$
$$= 2\pi$$
$$\text{Shaded area} = 2\pi - \frac{1}{2} \times x \times x$$
$$= 2\pi - \frac{1}{2} \times x^2$$
$$= 2\pi - 4$$

∴ the shaded area is $(2\pi - 4)$ cm^2.

4. **80 cm**

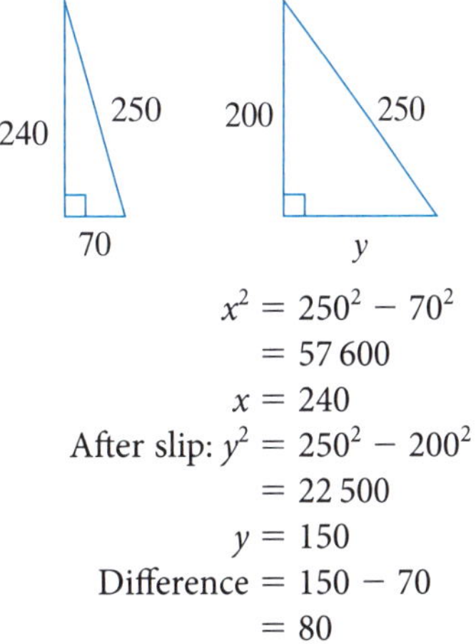

$$x^2 = 250^2 - 70^2$$
$$= 57\,600$$
$$x = 240$$
$$\text{After slip: } y^2 = 250^2 - 200^2$$
$$= 22\,500$$
$$y = 150$$
$$\text{Difference} = 150 - 70$$
$$= 80$$

∴ the ladder moves 80 cm.

5. 50 km

Yacht A: Distance $= 12 \times 2.5$
$= 30$

Yacht B: Distance $= 16 \times 2.5$
$= 40$

$$x^2 = 30^2 + 40^2$$
$$= 2500$$
$$x = 50$$

$\therefore$ the boats are 50 km apart.

6. $18(\pi - 2)$ cm^2

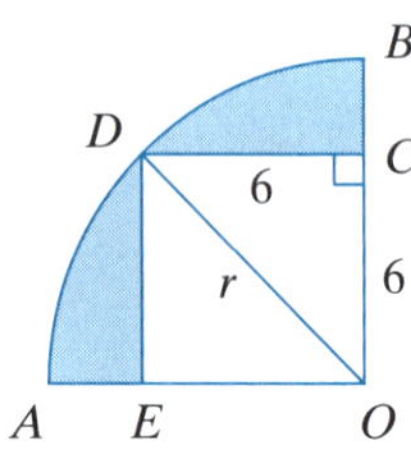

$$r^2 = 6^2 + 6^2$$
$$= 72$$

Shaded area = Area of quadrant − area of square

$$= \frac{1}{4}\pi r^2 - 6^2$$
$$= \frac{1}{4} \times \pi \times 72 - 36$$
$$= 18\pi - 36$$
$$= 18(\pi - 2)$$

$\therefore$ the area is $18(\pi - 2)$ cm^2.

Key Skill  45 Time (pages 110–111)

1. 11:50 pm Sunday

Time-difference: Perth is 15 h ahead of San Francisco.
1450 minus 15 h is 2350 previous day.
$\therefore$ the time is 11:50 pm Sunday in San Francisco.

2. 6:10 pm Wednesday

4:50 pm Tuesday plus 18 h is 10:50 am Wednesday. Add 20 min gives 11:10 am.
Plane arrives 11:10 am Wednesday (Rome local time).
Time-difference: Perth is 7 h ahead of Rome.
11:10 am plus 7 h is 6:10 pm.
$\therefore$ the time is 6:10 pm Wednesday in Perth.

3. 11:10 pm Saturday.

The table would show Sydney: '0630 Saturday'.
Time-difference: Sydney is 18 h ahead of San Francisco.
4:30 pm is 1630, minus 18 h gives 10:30 pm on the previous day. Adding 40 min gives 11:10 pm.
$\therefore$ the time is 11:10 pm Saturday in San Francisco.

4. 12:30 pm Friday

The table would show Chicago: '1430 Friday'.
Time-difference: Auckland is 18 h ahead of Chicago.
6:30 pm plus 18 h is 12:30 pm on the next day.
$\therefore$ the time is 12:30 pm Friday in Auckland.

5. 12:55 pm

The table would show New York: '1530 Friday'.
8:40 am plus 10 h is 6:40 pm. Add 15 min gives 6:55 pm.
Plane arrives 6:55 pm (New York local time).
Time-difference: Honolulu is 6 h behind New York.
6:55 pm minus 6 h is 12:55 pm.
$\therefore$ she arrives 12:55 pm Honolulu time.

6. 8:55 pm Tuesday

The table would show Adelaide: '0600 Saturday', and London '1930 Friday'.
Time-difference: Adelaide is 10 h 30 min ahead of London.
Conference: 9:45 am plus 40 min is 10:25 am.
10:25 am plus 10 h 30 min is 8:55 pm
$\therefore$ the conference finishes at 8:55 pm Tuesday in Adelaide.

Revision Test 13
Level of difficulty—Average (page 112)

1. 19.2 cm

$$\text{Vol. of cube} = 24^3$$
$$= 13\,824$$
$$\text{Vol. of rectangular prism} = lbh$$
$$13\,824 = 36 \times 20 \times h$$
$$720h = 13\,824$$
$$h = \frac{13\,824}{720}$$
$$= 19.2$$

$\therefore$ the height is 19.2 cm.

2. 157 cm

Diameter = 1.8 m, radius = 0.9 m
Also, 1 m^3 = 1 kL.

$$\text{Volume} = \pi r^2 h$$
$$4 = \pi \times 0.9^2 \times h$$
$$h = \frac{4}{0.81\pi}$$
$$= 1.571\,900\,673\ldots$$
$$= 1.57 \text{ (2 dec. pl.)}$$

$\therefore$ the height is 157 cm.

3. 56

Length: $28 \div 4 = 7$
Width: $16 \div 4 = 4$
Height: $10 \div 4 = 2$ with some remainder
$7 \times 4 \times 2 = 56$.
$\therefore$ Cheryl can fit 56 cubes.

4. $\frac{10}{\pi}$ **cm**

$$\begin{aligned}\text{Volume of cube} &= 10^3\\ &= 1000\\ \text{Volume of cylinder} &= \pi r^2 h\\ 1000 &= \pi \times 10^2 \times h\\ 100\pi h &= 1000\\ h &= \frac{1000}{100\pi}\\ &= \frac{10}{\pi}\end{aligned}$$

$\therefore$ the water is $\frac{10}{\pi}$ cm high.

5. **206 m**

$$\begin{aligned}x^2 &= 73^2 - 48^2\\ &= 3025\\ x &= \sqrt{3025}\\ &= 55\end{aligned}$$

Sides are 55 m and 48 m.

$$\begin{aligned}\text{Perimeter} &= 2(55 + 48)\\ &= 206\end{aligned}$$

$\therefore$ the perimeter is 206 m.

6. **10:20 pm**

Dunedin is 4 h ahead of Xi'an.
4:50 pm plus 1 h 30 min is 6:20 pm (Xi'an time).
6:20 pm plus 4 h is 10:20 pm.
$\therefore$ the time in Dunedin is 10:20 pm.

Revision Test 14
Level of difficulty—Challenging (page 113)

1. **5π cm**

Let $OD = x$.

$$\begin{aligned}x^2 &= 6^2 + 8^2\\ &= 100\\ x &= 10\end{aligned}$$

$\therefore$ the radius of the circle is 10 cm.

$$\begin{aligned}\text{Length of arc } AB &= \frac{1}{4} \times 2 \times \pi \times 10\\ &= 5\pi\end{aligned}$$

$\therefore$ the length is 5π cm.

2. **20 km**

$$\begin{aligned}x^2 &= 12^2 + 16^2\\ &= 400\\ x &= 20\end{aligned}$$

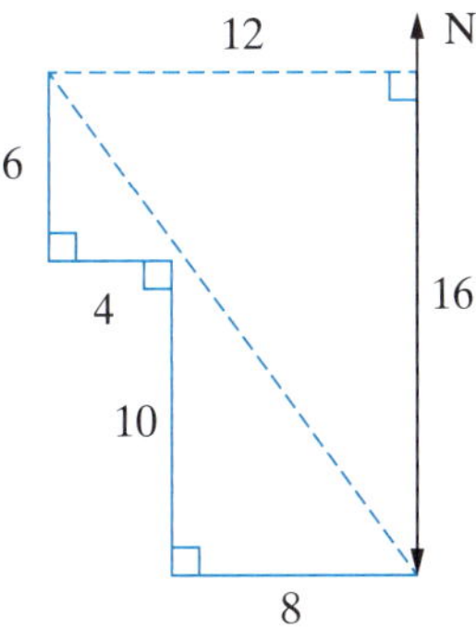

$\therefore$ the straight-line distance is 20 km.

3. **4:20 pm**

$8 - 0.4 = 7.6$
Water is filling at the rate of 7.6 L/min.

$$\begin{aligned}\text{Volume of container} &= 80 \times 70 \times 38\\ &= 212\,800\end{aligned}$$

$\therefore$ the volume is 212 800 cm^3.
$\therefore$ the capacity of the container is 212.8 L.

$$\begin{aligned}\text{Time to fill} &= 212.8 \div 7.6\\ &= 28\end{aligned}$$

$\therefore$ container is full at 7:28 am.

$$\begin{aligned}\text{Time to empty} &= 212.8 \div 0.4\\ &= 532\end{aligned}$$

$\therefore$ it takes 532 mins, or 8 h 52 min, to empty.
7:28 am plus 8 h is 3:28 pm plus 52 min is 4:20 pm.
$\therefore$ the container is empty at 4:20 pm.

4. **5:30 am Saturday**

$$\begin{aligned}\text{Time of flight} &= 11\,890 \div 820\\ &= 14.5\end{aligned}$$

9:00 pm Friday plus 14 h 30 min is 11:30 am Saturday (Brisbane time).
From table, Dubai is 6 h behind Brisbane.
11:30 am minus 6 h is 5:30 am.
$\therefore$ Sophie arrives at 5:30 am Saturday.

5. **4.5 cm**

$$\begin{aligned}\text{1st cylinder: Vol. of water} &= \pi \times 6^2 \times 8\\ &= 288\pi\\ \text{2nd cylinder: Vol. of water} &= \pi r^2 h\\ 288\pi &= \pi \times 8^2 \times h\\ h &= \frac{288\pi}{64\pi}\\ &= 4.5\end{aligned}$$

$\therefore$ the water is 4.5 cm deep.

6. **9 cm**

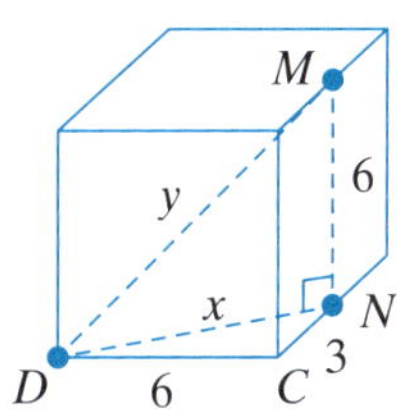

Locate N, the midpoint of CG, directly below M.

$$\begin{aligned}\text{Consider } \triangle DCN: x^2 &= 6^2 + 3^2\\ &= 45\\ \text{Consider } \triangle MND: y^2 &= 6^2 + 45\\ &= 81\\ y &= 9\end{aligned}$$

$\therefore$ the shortest distance is 9 cm.

STATISTICS AND PROBABILITY

Key Skill  Statistics A (pages 114–115)

1. **\$1470**

Mean of 6 salaries = 1295
Total of 6 salaries = 1295 × 6
= 7770
Total of 5 salaries
= 1160 + 1210 + 1290 + 1300 + 1340
= 6300
6th salary = 7770 − 6300
= 1470
∴ the salary is \$1470.

2. **3**

Mean of 8 games = 12
Total of 8 games = 12 × 8
= 96
Total of 9 games = 11 × 9
= 99
99 − 96 = 3
∴ Jerome scored 3 points.

3. **87**

Total of 4 assignments = 85 + 82 + 91 + 80
= 338
Total of 5 assignments = 85 × 5
= 425
Difference = 425 − 338
= 87
∴ Xui Li must score 87.

4. **32%**

Total of 4 tests = 83 × 4
= 332
If he scored 100% in each of 3 tests, he would score 32% in the other.
∴ the lowest possible score is 32%.

5. **97**

Total of 4 tests = 73 + 63 + 66 + 71
= 273
Total of 5 tests = 74 × 5
= 370
Difference = 370 − 273
= 97
∴ Belinda must score at least 97 in her final test.

6. **14**

Total of 8 games = 11.5 × 8
= 92
Total of 9 games = 92 + 4
= 96
Mean of 10 games = 11 × 10
= 110
Difference = 110 − 96
= 14
∴ Avery needs to score 14 points in the 10th game.

Key Skill  Statistics B (pages 116–117)

1. **8**

Range = 10 − 3
= 7
Scores are 3, 5, 6, x, 9, 10.
As range is 7, then $x = 8$.
∴ $x = 8$.

2. $\boldsymbol{a = 10, b = 1, c = 2, d = 3}$

If a total of 30 students, then $a = 10$.
∴ mode has 10 scores, ∴ $c = 2$.
As the scores represent number of children at home, then the lowest score must be 1.
If range = 3 = median, then $b = 1$ and $d = 3$.
∴ $a = 10$, $b = 1$, $c = 2$, $d = 3$.

3. $\boldsymbol{x = 12, y = 13}$

Total of 8 scores = 8 × 10
= 80
∴ $x + y = 80 - 55$
= 25
Median of 11 is middle of 10 and x:
∴ $x = 12$
∴ $y = 25 - 12$
= 13
∴ $x = 12$, $y = 13$

4. **3, 5 and 6**

Let the scores be 2, a, b, c, 4, 4.
Consider possibilities:
- $a = b = c = 3$: 2, 3, 3, 3, 4, 4—no, as median ≠ range
- $a = 3$, $b = 5$, $c = 6$: 2, 3, 4, 4, 5, 6—yes, as range = median = mode = mean = 4

∴ 3, 5 and 6 goals scored in other games.

5. $\boldsymbol{a = 7, b = 3, c = 5, d = 8}$

If a total of 15 scores, then $a = 7$.
∴ mode has 7 scores,
∴ $d = 8$.
As the range is 5, then $b = 3$.
As there are 15 scores, the middle score is c.
∴ $c = 5$.
∴ $a = 7$, $b = 3$, $c = 5$, $d = 8$.

6. 67

Let Kelvin's marks be a, b, 83, 84, 84.

$$\text{Total of 5 tests} = 5 \times 80 = 400$$

$$a + b = 400 - (83 + 84 + 84) = 149$$

If a and b are less than 83, then let $b = 82$.

$149 - 82 = 67$

The lowest mark is 67: i.e. 67, 82, 83, 84, 84

∴ lowest mark is 67.

Key Skill  Probability A

(pages 118–119)

1. $\mathbf{\frac{3}{5}}$

$$\text{P(yellow)} = \frac{6}{10} = \frac{3}{5}$$

∴ the probability of a yellow ball is $\frac{3}{5}$.

2. $\mathbf{\frac{13}{20}}$

Multiples of 2 or 3: 2, 3, 4, 6, 8, 9, 10, 12, 14, 15, 16, 18, 20

$$\text{P(multiple of 2 or 3)} = \frac{13}{20}$$

∴ the probability of a multiple of 2 or 3 is $\frac{13}{20}$.

3. $\mathbf{\frac{2}{3}}$

$$\text{P(odd)} = \frac{4}{6} = \frac{2}{3}$$

∴ the probability of an odd number is $\frac{2}{3}$.

4. $\mathbf{\frac{1}{3}}$

$$\text{P(neither red nor blue)} = \text{P(green)} = \frac{8}{24} = \frac{1}{3}$$

∴ the probability of not red or blue is $\frac{1}{3}$.

5. $\mathbf{\frac{5}{8}}$

$$\text{Year 11 cars} = 4$$

$$\text{Year 12 cars} = 4 \times 5 = 20$$

$$\text{Staff cars} = 64 - 24 = 40$$

$$\text{P(staff)} = \frac{40}{64} = \frac{5}{8}$$

∴ the probability that the car is driven by a staff member is $\frac{5}{8}$.

6. 32

$\text{P(male)} = \frac{3}{7}$ and $\text{P(female)} = \frac{4}{7}$.

∴ one-seventh of club members = 8

Seven-sevenths of club members = 8×7 = 56

There are 56 members in the club.

$$\text{Number of females} = \frac{4}{7} \times 56 = 32$$

∴ there are 32 female members of the club.

Key Skill  Probability B

(pages 120–121)

1. $\mathbf{\frac{1}{10}}$

Let P(Dianne wins) be x.

∴ P(Julian wins) is $3x$ and P(Kayla wins) is $6x$.

Sum of probabilities = 1

$$x + 3x + 6x = 10x = 1$$

$$x = \frac{1}{10}$$

∴ the probability that Dianne wins is $\frac{1}{10}$.

2. $\mathbf{\frac{4}{13}}$

Let the number of pink balls be x.

∴ there are $4x$ blue balls and $8x$ red balls.

$$x + 4x + 8x = 13x$$

$$\text{P(blue)} = \frac{4x}{13x} = \frac{4}{13}$$

∴ the probability of a blue ball is $\frac{4}{13}$.

3. $\mathbf{\frac{2}{5}}$

Let P(blue) be x.

∴ P(green) is x, P(yellow) is x and P(red) is $2x$.

Sum of probabilities = 1

$$x + x + x + 2x = 5x = 1$$

$$x = \frac{1}{5}$$

$$2 \times \frac{1}{5} = \frac{2}{5}.$$

∴ the probability of choosing a red is $\frac{2}{5}$.

4. $\mathbf{\frac{3}{10}}$

Let P(any player except George winning) be x.

∴ P(George winning) is $3x$.

Sum of probabilities = 1

$$7x + 3x = 10x = 1$$

$$x = \frac{1}{10}$$

$$3 \times \frac{1}{10} = \frac{3}{10}$$

∴ the probability of George winning is $\frac{3}{10}$.

5. $\frac{4}{15}$

Let P(David wins) be x.

$\therefore$ P(Sam wins) is $2x$, P(Matt wins) is $4x$, P(Joey wins) is $8x$.

Sum of probabilities = 1

$$x + 2x + 4x + 8x = 15x = 1$$

$$x = \frac{1}{15}$$

$$4 \times \frac{1}{15} = \frac{4}{15}.$$

$\therefore$ the probability of Matt winning is $\frac{4}{15}$.

6. $\frac{19}{34}$

Let the number of black frogs be x.

$\therefore$ there are $2x$ green frogs and $4x$ red frogs.

$$\text{Total frogs} = x + 2x + 4x$$
$$= 7x$$
$$34 + 1 = 35$$
$$7x = 35$$
$$x = 5$$

Orignally there were 5 black frogs, 10 green frogs and 20 red frogs (now 19 red frogs).

$$\text{P(red frog)} = \frac{19}{34}$$

$\therefore$ the probability of another red frog is $\frac{19}{34}$.

Key Skill  50 Probability C

(pages 122–123)

1. $\frac{33}{100}$

Both sports = 100 − (21 + 19 + 27)
= 33

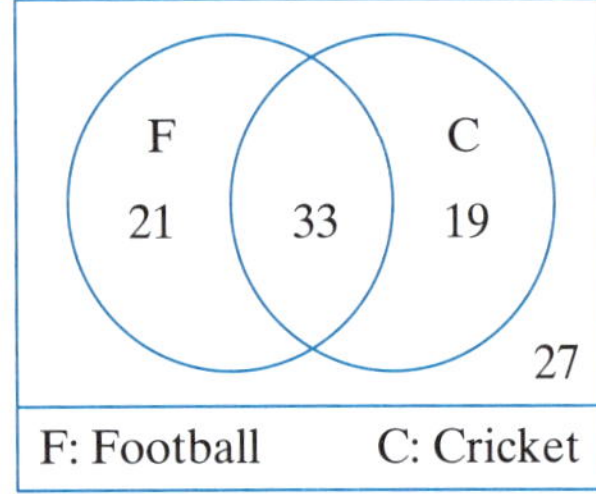

$$\text{P(both sports)} = \frac{33}{100}$$

$\therefore$ the probability of playing both sports is $\frac{33}{100}$.

2. $\frac{25}{81}$

Customer satisfaction survey			
	Pleased	Disappointed	Total
Lunch	28	10	38
Dinner	25	18	43
Totals	53	28	81

$$\text{P(pleased dinner guest)} = \frac{25}{81}$$

$\therefore$ the probability of a dinner guest who was pleased is $\frac{25}{81}$.

3. 20%

Only Jaycee = 250 − 190
= 60

Only UShop = 250 − 220
= 30

Neither shop = 350 − (60 + 190 + 30)
= 70

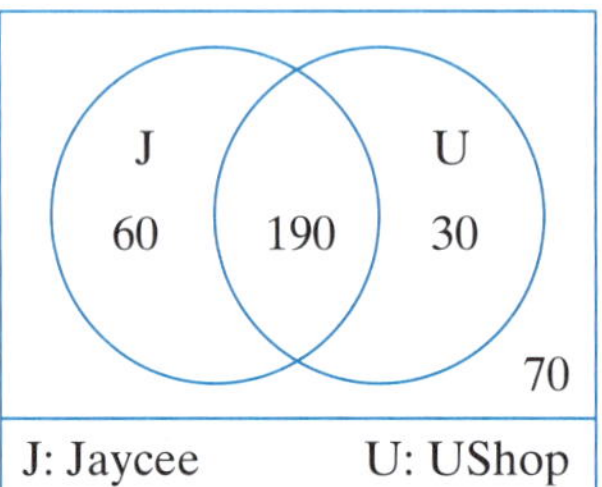

$$\text{P(neither store)} = \frac{70}{350}$$
$$= \frac{1}{5}$$

$$\frac{1}{5} \times 100\% = 20\%$$

$\therefore$ the probability of not shopping at either store is 20%.

4. 6

Students who received Grades A & B			
	No A's	At least one A	Totals
No B's	6	4	10
At least one B	8	6	14
Totals	14	10	24

Total with at least one A or B = 4 + 8 + 6
= 18

Three-quarters of class = 18

One-quarter of class = 6

$\therefore$ 6 students did not receive an A or a B.

5. $\frac{5}{12}$

Language of Chinese speakers		
	Mandarin	Not Mandarin
Cantonese	$\frac{1}{4}$	$\frac{1}{4}$
Not Cantonese	$\frac{5}{12}$	$\frac{1}{12}$

$$\text{Cantonese only} = \frac{1}{2} \text{ of } \frac{1}{2}$$
$$= \frac{1}{4}$$

Mandarin only $= 1 - (\frac{1}{4} + \frac{1}{4} + \frac{1}{12})$

$= \frac{5}{12}$

$\therefore$ the probability they only speak Mandarin is $\frac{5}{12}$.

6. $\frac{1}{3}$

Tea only $= 0.2 \times 60$
$= 12$

Coffee only $= 0.3 \times 60$
$= 18$

Remaining people $= 60 - (12 + 18)$
$= 30$

$\therefore$ 3 parts $= 30$

$\frac{1}{3} \times 30 = 10$

$\frac{2}{3} \times 30 = 20$

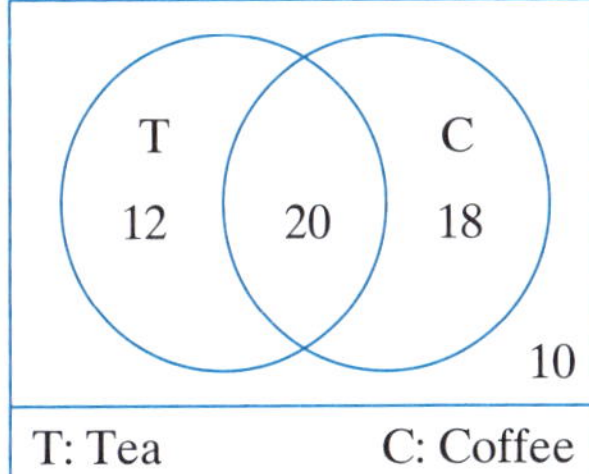

P(both tea and coffee) $= \frac{20}{60}$

$= \frac{1}{3}$

$\therefore$ the probability of the person drinking both is $\frac{1}{3}$.

Revision Test 15
Level of difficulty—Average

(page 124)

1. **8 and 12**
If the median is 6, then $x = 8$.
If the range is 9, then $y = 12$.
The children's ages are 3, 4, 8, 12.
$\therefore$ the ages are 8 and 12.

2. $\frac{1}{3}$
The bag now contains 4 green and 2 yellow.

P(yellow) $= \frac{2}{6}$

$= \frac{1}{3}$

$\therefore$ the probability of a yellow gumdrop is $\frac{1}{3}$.

3. $\frac{1}{10}$
Less than 16, odd, composite: 9, 15.

P(< 16, odd, composite) $= \frac{2}{20}$

$= \frac{1}{10}$

$\therefore$ the probability is $\frac{1}{10}$.

4. $\frac{2}{3}$
Even or less than 3: 1, 2, 4, 6

P(even or less than 3) $= \frac{4}{6}$

$= \frac{2}{3}$

$\therefore$ the probability that it is even or less than 3 is $\frac{2}{3}$.

5. **12 kg**

Total mass of 3 dogs $= 14 \times 3$
$= 42$

Bella's mass $= \frac{42 - 18}{2}$
$= 12$

$\therefore$ Bella's mass is 12 kg.

6. $\frac{1}{30}$

Only vinegar $= 60 - (16 + 30 + 12)$
$= 2$

$\therefore$ 2 people wanted vinegar and no salt.

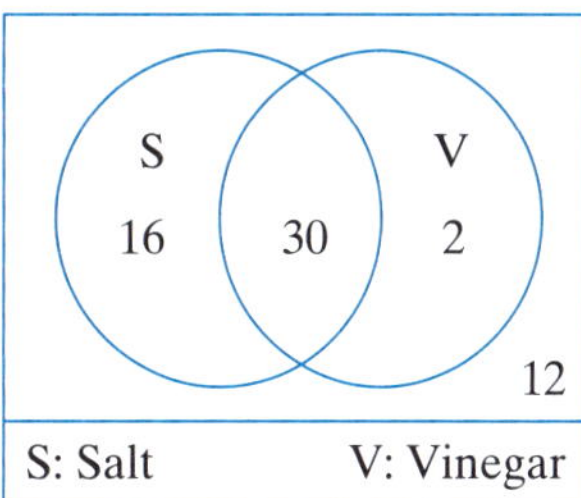

P(vinegar and no salt) $= \frac{2}{60}$

$= \frac{1}{30}$

$\therefore$ the probability the customer wanted vinegar without salt was $\frac{1}{30}$.

Revision Test 16
Level of difficulty—Challenging (page 125)

1. **18**

$$\begin{aligned}\text{Total of four tests} &= 12 + 16 + 14 + 10\\ &= 52\\ \text{Mean of four tests} &= 52 \div 4\\ &= 13\\ \text{Mean of five tests} &= 14\\ \text{Total of five tests} &= 14 \times 5\\ &= 70\\ \text{Difference} &= 70 - 52\\ &= 18\end{aligned}$$

$\therefore$ George needs 18 on his next test.

2. $\mathbf{\frac{3}{17}}$

Let the probability that Cody wins be $2x$.
$\therefore$ the probability that Winston wins is $6x$, that Sam wins is $6x$ and that Oscar wins is $3x$.
Sum of probabilities = 1

$$2x + 6x + 6x + 3x = 17x = 1$$
$$x = \frac{1}{17}$$
$$3 \times \frac{1}{17} = \frac{3}{17}$$

$\therefore$ the probability of Oscar winning is $\frac{3}{17}$.

3. $\mathbf{\frac{2}{3}}$

Composite are 21, 22, 24, 25, 26, 27, 28.
Odd composite balls are numbered 21, 25, 27.
P(multiple of 3 from the odd composites) = $\frac{2}{3}$.
$\therefore$ the probability is $\frac{2}{3}$.

4. **8 and 10.**

The mode is 4. This means the median is 8.
$\therefore x = 8$.
The range is 20. This means the mean is 10.
Sum of the scores = 50

$$\begin{aligned}4 + 4 + 8 + y + 24 &= 50\\ y + 40 &= 50\\ y &= 10\end{aligned}$$

$\therefore$ the missing numbers are 8 and 10.

5. $\mathbf{\frac{3}{20}}$

Overseas travellers survey			
	Overseas	Not overseas	Total
Males	5	3	8
Females	8	4	12
Total	13	7	20

There were 3 males who had not travelled overseas.
$\therefore$ the probability is $\frac{3}{20}$.

6. **30**

First, identify the percentages for the Venn diagram.
$15 + 15 + 10 = 40$

$$\begin{aligned}60\% \text{ of students} &= 72\\ 15\% \text{ of students} &= 72 \div 60 \times 15\\ &= 18\\ 10\% \text{ of students} &= 72 \div 6\\ &= 12\end{aligned}$$

The numbers for the Venn diagram can now be recorded.

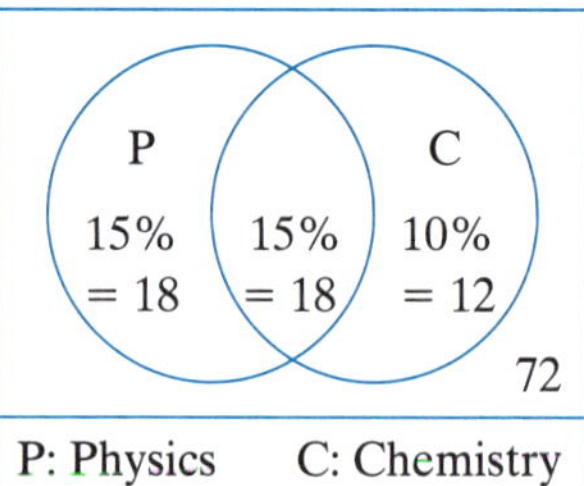

There are 120 students in Year 12.
$18 + 12 = 30$
$\therefore$ there are 30 students studying exactly one of these subjects.